ANALOG CIRCUITS AND SIGNAL PROCESSING

Series Editors
Mohammed Ismail, The Ohio State University
Mohamad Sawan, École Polytechnique de Montréal

For further volumes:
http://www.springer.com/series/7381

Nianxiong Nick Tan • Dongmei Li • Zhihua Wang
Editors

Ultra-Low Power Integrated Circuit Design

Circuits, Systems, and Applications

 Springer

Editors
Nianxiong Nick Tan
Zhejiang University,
Hangzhou
China

Zhihua Wang
Tsinghua University
Beijing
China

Dongmei Li
Tsinghua University
Beijing
China

ISSN 1872-082X
ISBN 978-1-4419-9972-6
DOI 10.1007/978-1-4419-9973-3
Springer New York Heidelberg Dordrecht London

ISSN 2197-1854 (electronic)
ISBN 978-1-4419-9973-3 (eBook)

Library of Congress Control Number: 2013951809

Contents

Contributors

Liming Chen ASIC and System Department, Institute of Microelectronics, Chinese Academy of Sciences, Beijing, China

Xinkai Chen Ecore Technologies Ltd., Beijing, China

Baoyong Chi Institute of Microelectronics, Tsinghua University, Beijing, China

Yong Hei ASIC and System Department, Institute of Microelectronics, Chinese Academy of Sciences, Beijing, China

Hanjun Jiang Institute of Microelectronics, Tsinghua University, Beijing, China

Xueping Jiang Smart Grid Research Institute, State Grid Corporation of China, Beijing, China

Quan Kong Vango Technologies, Inc., Hangzhou, China

Dongmei Li Department of Electronic Engineering, Tsinghua University, Beijing, China

Fule Li Institute of Microelectronics, Tsinghua University, Beijing, China

Liyuan Liu Institute of Semiconductors, Chinese Academy of Sciences, Beijing, China

Changyou Men Vango Technologies, Inc., Hangzhou, China

Nianxiong Nick Tan Institute of VLSI Design, Zhejiang University, Hangzhou, China

Zhihua Wang Institute of Microelectronics, Tsinghua University, Beijing, China

Nanjian Wu Institute of Semiconductors, Chinese Academy of Sciences, Beijing, China

Jinyong Xue ASIC and System Department, Institute of Microelectronics, Chinese Academy of Sciences, Beijing, China

Kun Yang Vango Technologies, Inc., Hangzhou, China

Zenghui Yu ASIC and System Department, Institute of Microelectronics, Chinese Academy of Sciences, Beijing, China

Jia Yuan ASIC and System Department, Institute of Microelectronics, Chinese Academy of Sciences, Beijing, China

Lingwei Zhang Galaxycore Shanghai Ltd. Corp., Shanghai, China

Yan Zhao Vango Technologies, Inc., Hangzhou, China

Shupeng Zhong Vango Technologies, Inc., Hangzhou, China

Introduction

Xueping Jiang and Nianxiong Nick Tan

We introduce the emerging applications for ultra low power devices in this chapter. Moore's law continues to drive function integration. The transistor density is defined as the transistor number on one square silicon die that is generally considered the largest manufacturable die. Such a die can now hold over 7 billion transistors. It is about the same as the population of the planet. But whether those transistors are active or not, they consume power. To satisfy Moore's law, die size has been increasing at the rate of 7% per year. The operating frequency has doubled every two years. Therefore to meet the performance goal, the supply voltage scales by only ~15% every two years, rather than the theoretical 30%. Reduction in power consumption is not tracking Moore's Law. Additionally, at smaller geometries the cost of the fabrication is growing substantially, which affects wafer pricing. Hence both power reduction and cost reduction are not tracking Moore's Law. This inspires creative thinking and promotes research and development in low power circuits and architecture. So that raises some interesting challenges about power, cost, and performance that need innovative solutions.

System-on-chip (SoC) has become a reality since increasing complexity is driving new business models. SoC usually integrates all components of a computer or other electronic system into a single chip. The key electronic chip we focus on is a "system-on-chip", which is a must for most modern sophisticated devices. It is critical for chip designers to work directly with system designers or customers in product definition in order to meet the design requirements for power, cost, and performance.

In particular, we focus on short-channel effects, device parameter variations, excessive junction and gate oxide leakage, as the main obstacles dictated by fundamental device physics. Functionality of special circuits in the presence of high leakage, SRAM cell stability, bit line delay scaling, and power consumption in clocks

X. Jiang (✉)
Smart Grid Research Institute, State Grid Corporation of China, Beijing, China
e-mail: xpj@yahoo.com

N. N. Tan
Institute of VLSI Design, Zhejiang University, Hangzhou, China

N. N. Tan et al. (eds.), *Ultra-Low Power Integrated Circuit Design*,
Analog Circuits and Signal Processing 85, DOI 10.1007/978-1-4419-9973-3_1,
© Springer Science+Business Media New York 2014

and interconnects, are the primary design challenges. Soft error rate control and power delivery pose additional challenges. All of these problems are further compounded by the rapidly escalating complexity of SoC designs. The excessive leakage problem is particularly severe for battery-operated, high-performance SoCs.

1 Emerging Applications

The electronics industry has made spectacular progress since the invention of the integrated circuit in 1958. Discrete devices have been replaced by integrated circuits. In more than 50 years, the technology has rapidly evolved from producing simple chips with several components to fabricating very complicated chips containing billions of transistors. The number of transistors per chip has continued to increase exponentially over the years, which makes feasible more and more functions to be integrated into one chip. At the same time, the minimum feature size of transistors has dropped exponentially accordingly, which reduces both silicon area and cost per function. Most of the research and development efforts have been oriented towards increasing operation speed and complexity. This gives rise to the so-called high performance-to-cost ratio. While paying attention to speed and area, power consumption has been ignored or often considered as of minor concern.

The scenario is, however, undergoing some radical changes. Integrated circuits have incorporated more and more functions on each chip until now it is essentially possible to put an electronic system into a single silicon chip. This chip is termed as a SoC that usually integrates all components of a computer or other electronic system into a single chip. It may consist of digital, analog, and sometimes radio-frequency (RF) functions on a single chip substrate.

The motivation for ultra low power electronics has stemmed from two major requirements: (1) for portable or wearable implantable battery operated systems that are sufficiently small in weight and volume but require long battery operation time; (2) for batteryless systems in which energy is harvested from environment. For example, the chips in various sensor wireless networks can operate with energy harvested from environment, which will be widely used in the internet of thing (IoT) such as smart grid.

Nowadays, SoCs have obtained more and more functionalities. They have become smarter than ever before. Thanks to the rapid development of IC manufacture industry, on any of today's SoCs, extraordinary high dense transistors are integrated. They are the key devices or chips in systems such as portable or wearable medical electron devices, smart phones, wireless smart terminals, multi-media terminals, wireless sensor networks, wearable health care monitoring devices and so on. However, problems accompanying such high density of integration emerge such as dynamic power, leakage power, thermal issues, and limited battery operation hours. In most consumer electronic and bio-medical applications, devices are required to be portable, wearable or even implantable. Hence, the sizes of the systems are forced to be small. As a result compact batteries are required as energy sources for reasonable operation time. Due to limited energy storage capacity of batteries, in order to guarantee a reasonable battery operation time, various low power design techniques are greatly demanded to make a drastic reduction of the power consumption.

Endoscopic capsule is such a kind of application that allows people to directly examine the entire intestine and does not require any sedation, anesthesia or insufflations of the bowel. Such noninvasiveness of medical examination procedure also improves patients' comfort. The small-sized capsule that can be swallowed by an adult is a very complex system in which, a LED, a CMOS image sensor, a SoC and an antenna for wireless transmission are included. The average life time for the capsule is 8 h during which the capsule can travel through the intestine and then evacuate from the human body. The problem is that the evacuation time for various persons is quite different. For this reason we should extend the actual battery operation time with the SoC running inside the capsule to more than 16 h in order to cover a wide range of evacuation time. As a result various low power design techniques are necessary in the capsule SoC.

Another application in which low power SoC is needed is the digital hearing aid system. The purpose of the hearing aid is to amplify sound signal in order to compensate the hearing loss of patients. Compared to the conventional amplifier that provides constant gain for all frequency components, the hearing aid needs to provide different gain for signal in various frequency sub-bands. That's because sensitivity of human ears vary greatly with sound frequencies. Demands for digital hearing aids and other portable digital audio devices are becoming more and more challenging: On the one hand more computational performance is needed for new algorithms including noise reduction, improved speech intelligibility, beam forming, etc. On the other hand flexibility through programmability is needed to allow for product differentiation and longer lifetime of hardware designs. Both requirements have to be met by ultra-low power solutions being operated out of very small batteries. A hearing aid is a quite small device that can be worn behind the ear, fitting entirely in the outer ear or even put inside the canal. In such a small system, a microphone, an amplifier and an output transducer are required to perform sound signal acquisitions, sound amplification and playback, respectively. Recently, digital hearing aid dominates the market. The traditional amplifier is already replaced by the SoC with millions of transistors. The sound signal is firstly acquired and digitized and then processed by a digital signal processor. Sound is amplified in the digital domain. Other advanced functions can also be performed such as directional dependent amplification, echo cancellation and noise suppression or sound classification. Usually the Zinc-air battery is employed to power the whole system due to its relatively larger energy density with small size. However owing to the limit size, the energy capacity is still small. A size-10 battery can provide a current of 1 mA for a total 90 operation hours or 1.5 mA for 60 operation hours. Suppose an 8 h operation time per day, such energy storage is only available for less than two weeks. To prevent frequent battery replacement the SoC inside the hearing aid system must have ultra low power consumptions. Hence, new hearing aids are examples of ultra low power designs and technologies.

Even in energy metering applications, ultra-low power is desirable. First of all, the energy meter is always on. The power dissipated by the energy meter is the power wasted. To prevent tampering and maintain the real time clock (RTC), most energy meters have a battery. It is required that the meters can be in stand-by mode for more than 5 years. In the stand-by mode, some circuits such as the RTC and monitoring circuits have to be active. In certain anti-tampering applications, the meter needs to wake up periodically on the battery to detect or meter, thus ultra-low power is desired. Another requirement of ultra low power is the consideration of the power supply to the metering chip. Using potential transformers (PTs) have drawbacks such as being bulky, leaky, and poor reliability. Use a RC supply can overcome the problems associated with PT based supplies. However, for RC supplies, the smaller the capacitance, the better reliability and lower cost. With smaller capacitance, the RC supply cannot provide so much current, requiring ultra-low power metering chips.

In conclusion, due to its enabling potential for new markets and its importance for competitiveness, the further reduction of the electric power for SoCs and embedded systems has recently become one of the most challenging areas of research and development in this industry. In order to obtain the ultra-low power for these products, the different aspects of ultra-low power design for ASICs and embedded systems at all levels of abstractions from system level to circuit level and technology should be considered.

2 Design Challenges for Ultra-Low Power IC

Power consumption for ICs is as important as performance. The motivation for ultra-low power ICs is portability where battery operation life, increased functionality and heat generation are the key factors. The other application is huge server farms or large volume of devices such as energy metering in smart grids due to environmental awareness. Usually, the amount of energy stored in a battery is directly proportional to the battery weight and size. A battery for portable devices with limited weight and size can store limited amount energy. When a SoC is operated with a battery as a supply, the battery operation time operating the SoC is determined by the ratio of the energy stored in the battery to the average SoC power consumption. Since the battery can be seen as an ideal voltage source with constant output voltage, the power provided by the battery is then directly proportional to the current drawn by the SoC. In order to obtain longer battery operation time, the average SoC power consumption has to be reduced. In Fig. 1, we show a CMOS inverter as an example to explain power dissipation by digital circuits.

It is well known that for digital CMOS circuits, average dynamic current drain can be expressed as:

$$I = \alpha C_{load} V_{dd} f_{clk} \tag{1}$$

Fig. 1 Dynamic CMOS
switching

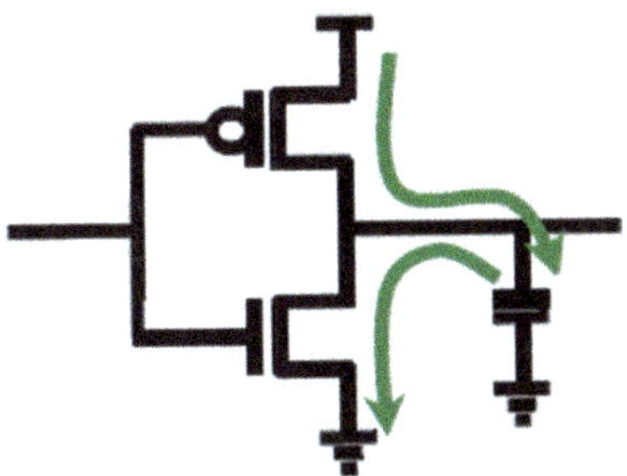

In the above expressions, α represents the average activity factor of a digital system. C_{load} is the total effective load capacitance. V_{dd} is the supply voltage and f_{clk} represents the system operating frequency. Dynamic power consumes when the output of a gate changes states. Since not all the gates switch at the same time, the term α is introduced to model such effect.

It gives rise to the average dynamic power consumption as follows.

$$P = \alpha C_{load} V_{dd}^2 f_{clk} \qquad (2)$$

In the above expressions, dynamic power consumption is quadratically dependent on supply voltage.

Clearly there are two ways for reducing dynamic power in CMOS digital circuits: (1) supply voltage reduction, and (2) activity reduction. Since power consumption depends quadratically on supply voltage, supply voltage reduction can give rise to substantial power saving. There are two schemes to employ supply voltage reduction without compromising performance: multiple power domains, and dynamic frequency and voltage scaling (DFVS).

To reduce the activity factor in CMOS digital circuits, there are three schemes to use for power consumption reduction: clock gating, power gating, or architectural exploration.

There is a design trade-off between chip frequency and voltage. Lowering the voltage will increase delay and reduce highest achievable operating frequency. The other thing we need to differentiate is power and energy. Energy is a time integral of power. With fixed energy, the power can be small with long activity time or the power can be large with short activity time. In the self-harvesting wireless sensing network for example, very small power from environment is harvested, conditioned and stored, the wireless transmitter can transmit data with high power of short period time. Usually the operation duty cycle for sensing wireless network will be less than 1%.

Power consumption is difficult to estimate. It depends on number of elements, activity factor of each element, specific cells used (size, shape, technology etc.), manufacturing variance, operating temperature. The existing tools use measurements from previous designs, power models, or complicated formulae to estimate power consumption. It is a must to verify functional correctness of circuits that

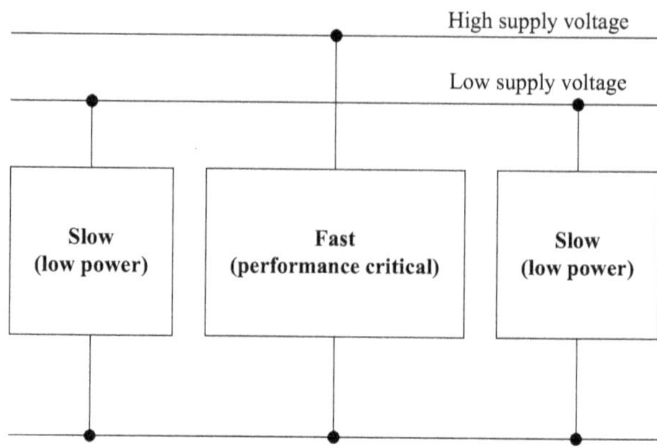

Fig. 2 Multiple power domains

employ low power design techniques. Currently, power saving techniques contain (1) multiple power domains; (2) dynamic frequency and voltage scaling; (3) clock gating; (4) power gating; (5) architectural exploration, etc. Such a multiple power domain is illustrated in the following figure.

In the scheme of multiple power domains as shown in Fig. 2, different blocks operate with different voltages and frequencies to achieve the optimized tradeoff between power and performance. Low power blocks can operate at lower voltage and slower frequency as long as performance meets the requirements. Higher supply voltage is used for performance critical logic, which runs at higher speeds, and consumes higher power. Multiple power domains have asynchronous interfaces for data and clock between power domains. It is necessary to model the data and clock correctly for function and performance verification. When a logic signal from the slow block with the lower supply voltage is connected to the fast block with the higher supply voltage, the signal level is closer to threshold voltage of the transistors, and can consume excessive leakage power and reduce noise margin. This scheme also requires additional power supply grid, and associated support such as decoupling capacitors, to ensure error-free operation.

In the scheme of dynamic frequency and voltage scaling (DFVS) method, power consumption reduction is achieved by dynamically controlling the voltage and the clock frequency. The digital chip is designed to deliver maximum performance at the highest supply voltage. The chip is operated at increased voltage when a higher performance is required. Otherwise, the chip is operated at a lower voltage with substantial power reduction when low power is demanded. On-chip power management circuitry dynamically controls the supply voltage and the clock frequency. The digital chip can detect the performance demand and adjust the clock frequency and the supply voltage accordingly. Special attention should be paid to guarantee that there is no effect on functionality when the supply voltage adjusts according to the tradeoff between power and performance. A well designed digital SoC detects peaks in performance demand, adjusts the supply voltage and clock frequency to deliver the necessary throughput only when needed, therefore considerable saving in power and energy can be achieved.

Table 1 Summary of low-power design techniques

Power-save technique	Benefit	Timing penalty	Area penalty
Clock gating	Medium	Little	Little
Multiple voltage domains	High	Some	Little
Dynamic frequency and voltage scaling (DFVS)	High	Some	Some
Power gating	Huge	Some	Some

In the scheme of the clock gating, effectively preventing the clock from ticking can save dynamic power. The clock signal to a digital block is gated by a control signal, inhibiting the clock when the digital block is not in use, therefore reducing the clock signal activity and the overall dynamic power consumption. The clock gating can be inside latches and on the clock distribution network. In most standard cell libraries, there are special purpose cells that perform the function with a bit cost of area. It is much more effective to reduce power if the same gating function is applied to large sets of registers or the clock distribution network. Approximation of gating functions enables gating more registers with the same function. For example, the register signal c is used to select signals a or b. Then signal c can be used as the gating function for a and b. there are tools to automatically apply gating at the netlist level. Hand crafted clock-gating can be more effective for experienced engineers. Since clock accounts for substantial activity in the digital block, the clock gating results in considerable power saving. The clock gating is reported to reduce power consumption by 20–60 %. This techniques is not limited to clock signals alone but can be used for other signals having high activity. One has to be careful, however, to ensure that the power consumed in the digital block to detect and disable clock gating is considerably lower than the power reduction achieved by the clock gating.

Power gating completely shuts off the power to parts of the design when they are not being used. The virtual power and ground grids need careful design to ensure that logic state is not lost when the virtual supplies collapse. A powered-off gates must not drive powered-on gates since powered-off flip-flops lose their states. They will have arbitrary values when they are turned on. It is necessary to copy the reserved state back when power is restored. In order to avoid the arbitrary values of flip-flops (X value), a power management is needed to control power-off and power-on sequences.

The highest potential for reducing power consumption is at the system architecture level. There are many factors involved such as system partitioning, bandwidth on buses, pipelining, redundancy, and performance-critical blocks, etc. System architects need to make tradeoff among power, cost, and performance. Unfortunately the current power estimation tools are not accurate, which make the process less reliable (Table 1).

In addition to dynamic power consumption, there is another category of power that is the leakage power. It is consumed by each element at all times. It grows exponentially when voltage is reduced as technology shrinks. It increases as transistor size shrinks. Until recently, leakage current was negligible. However, leakage power is now a substantial component of the total power for the advanced technology.

The supply voltage continues to decrease with each technology generation. In order to improve transistor and circuit performance per technology generation, the power supply V_{dd} and transistor threshold voltage V_{th} should reduce at the same rate. However, reduction in V_{th} causes transistor sub-threshold leakage current (I_{off}) to increase exponentially. Notice that in some condition almost half of the power consumption may be from sub-threshold leakage.

Dual or triple V_{th} design technique is widely used to reduce the sub-threshold leakage power. In this technique, the process technology provides two or three flavors of transistors: high threshold voltage (high V_{th}), standard threshold voltage (standard V_{th}), and low threshold voltage (low V_{th}). The high V_{th} transistors yield slower logic, but lower leakage, whereas low V_{th} transistors yield faster logic, but higher ($\sim 10X$) leakage.

Here is the multiple V_{th} low power design methodology for a typical logic block using path delay distribution. If we use high V_{th} transistors everywhere, then it yields low leakage current, higher delay, or lower frequency. On the other hand, if we use low V_{th} transistors exclusively, then it yields higher frequency, but $\sim 10X$ higher leakage power. A selective insertion of low V_{th} transistors for performance critical circuits yields higher frequency with lower leakage power.

Coming with nanometer-scale process technology and green-movement of consumer electronics, the methodology of low power-oriented design takes place in the cutting edge implementation needs. Multi-supply Multi-voltage (MSMV) design approach accommodates multiple power domains and multiple voltage levels, which makes ultra-low power chip design much more complicated than the design of a normal chip. The challenges that MSMV implementation design needs to overcome in advanced process nodes: timing, power, design for testing (DFT), design for manufacture (DFM), design for yield (DFY), etc. MSMV scheme of ultra low power solution has the following features: multiple power domains inside, multiple power supplies for chip core cell, multi-V_{th} optimization, level shifter, isolation cell, retention cell, power gating cell, etc.

MSMV solution meets the most up-to-date demand of ultra low power for consumer electronics, such as: mobile internet device, smart phone, GPS, personal media player, other mobile/handheld devices, energy harvesting devices, etc. Comparing two similar SoCs with only one difference, one with MSMV design and the other without it, the power consumption of the SoC with MSMV could be greatly reduced to one-fiftieth (1/50) maximum of the one without MSMV in leakage power. As for dynamic power, it saves up to 70% power consumption in average under normal operation conditions.

The progress of the CMOS technology allows capacitance and supply voltage shrinking down aggressively. For 65 nm general purpose technology, the supply voltage is only 1 V. According to the prediction of ITRS, in 2020, the supply will lower down to only 0.6 V. Although the dynamic power dissipation shown above gets smaller along with the supply voltage, in some specific applications such as endoscopy capsule, digital hearing aids, wireless sensor network nodes, the power consumption is still not very small due to complex functionalities which SoCs perform. Various low power techniques on circuits design are still needed. As we discussed before, attractive techniques available are power gating, clock gating, multi-voltage/multi-frequency, and parallel processing, but exploring those techniques greatly increases the design effort. Thus new design methods, verification methods are sorely needed. In CMOS design, the leakage power that was assumed to be negligible is becoming more and more significant. The reason for that is that threshold voltage shrinks down much more slowly than that of supply voltage. So the transistors in CMOS logic cannot be completely turned off in the sub-threshold region. As a result the leakage current of a transistor increases drastically. Employing low power technology can greatly reduce the leakage at the cost of transistor performance degradation. Another significant source is the gate leakage current in nano-meter devices. The assumption that the MOS gate is high impedance is not accurate as the device geometry further shrinks.

Analog building blocks such as front-end programmable amplifier, analog-to-digital converter (ADC), bandgap reference, clock generator and low-drop out voltage regulator are necessary in most SoCs. Different analog and RF circuits have different design challenge due to their inherently different architectures. Although progressing of the CMOS technology brings benefits such as performance enhancement and power reduction to the digital design, analog and RF circuits design gets more difficult. Firstly the supply voltage drops while noise does not (more often the noise increases) and hence the dynamic range of analog signal reduces. The voltage headroom is also limited by the supply and swing gets smaller which results into significant distortion problems besides the degradation of signal-to-noise ratio (SNR). Employing operational amplifier and feedback is a popular method to obtain high precision signal conditioning. However, with a low supply voltage, cascode technique is no longer suitable. If cascade is used to increase the gain, more branches between power and ground are needed. More power consumption is required to push poles far away in order that the amplifier is stable in feedback configurations. Another problem associated with SoC is the noise generated by the switching of digital circuits. It can reduce the SNR of the analog circuits significantly. There is no universal way to reduce the power dissipation of analog and RF circuits. Sometimes the figure of merit (FOM) is used to quantify the SNR performance vs. bandwidth vs. power dissipation. In actual industrial designs, optimization is generally towards the whole system instead of merely looking at the FOM of a single block. Also looking at the FOM of a single block can be misleading if the driving circuits and to be driven circuits are not considered.

In this book, ultra low power design techniques are discussed and actual examples of industrial design are introduced. In Chap. 2 the interface to the real world is presented. An important conclusion from this chapter is that analog signal from a sensing network is usually weak and noisy. It needs to be processed before it can be used. Chapter 3 describe low noise low power amplifier. After flicker noise and thermal noise are introduced, the techniques to cancel offset and low frequency noise are presented. Noise reduction-related design examples are given at the end of this chapter. Chapter 4 gives an in-depth view of ultra-low power design principles for both ADC and DAC. Several ultra-low power design examples of ADC and DAC are also presented in the chapter. In Chap. 5, we introduce the ultra-low power digital circuit design. Sources of power dissipation in CMOS circuits are analyzed. The basic low power design methodologies are given. Application specific instruction-set processors (ASIP) are introduced and optimized for ultra low power audio application. The compromise between power and performance is explored. Chapter 6 discusses ultra-low power transceiver design. After general techniques for ultra-low power transceivers are described, several examples of transceiver designs are given. In the following chapters, real industrial design examples are presented. In Chap. 7, an ultra-low power energy metering chip is presented with the focus on both analog and digital circuit design. Chapter 8 presents an energy metering SoC that explores the system level power optimization as well as circuits-level optimization. Chapter 9 presents an ultra-low power hearing aids SoC and Chap. 10 presents an SoC for capsule endoscope. Both SoCs employ extensively ultra-low power techniques discussed in this book.

Interface to the Analog World

Liyuan Liu and Zhihua Wang

1 Sensoring the World

Sensors or detectors are ubiquitous in the world. Everyday millions of them are produced and integrated into various kinds of systems, e.g. TV set, refrigerator, smart phone, PC, laptop, elevators, automobile, plane and so on. With sensors electronic systems can see the world, feel the world and react with the world. With sensors electronic systems seems to be alive and they become much smarter than ever before which makes our daily life more convenient and more comfortable. Sensors are devices which can convert physical or chemical quantities, e.g. luminous intensity, temperature, air pressure, acceleration, height, distance, weight, sound pressure, PH value and so on into signals which can be measured, acquired, processed and stored easier. Nowadays such signals are always electric signals like voltage, current or frequency thanks to the development of modern microelectronics techniques.

Ideal sensors output signals which are proportional to the measurement or to some simple mathematical function of the measurement. The ratio between the output and the input is called sensitivity of the sensor. It means how large the output is if a unit quantity of the input signal is apply to the sensor. However real sensors usually have various non-idealities which makes their characteristics different from the ideal one. In order to characterize such deviations, some parameter of the sensors must be introduced.

- **Sensitivity error** is defined as deviation of sensitivity in practice from the value specified. Usually if the sensor is still linear, such error can be calibrated by simple mathematic manipulation in the post signal processing flow.

L. Liu (✉)
Institute of Semiconductors, Chinese Academy of Sciences, Beijing, China
e-mail: liuly@semi.ac.cn

Z. Wang
Institute of Microelectronics, Tsinghua University, Beijing, China

N. N. Tan et al. (eds.), *Ultra-Low Power Integrated Circuit Design*,
Analog Circuits and Signal Processing 85, DOI 10.1007/978-1-4419-9973-3_2,
© Springer Science+Business Media New York 2014

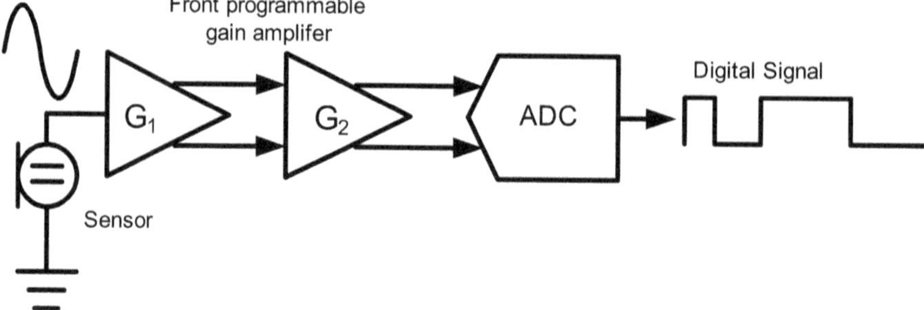

Fig. 1 Analog front end circuits for weak signal conditioning

- **Non-linearity** exists if the sensitivity is not constant and has relations to the input signal. This error can also be adjusted as long as the real characteristic of the sensor can be obtained.
- **Offset** is defined as the output value when zero input is applied to the sensor. Usually the offset is a constant value and can be eliminated by the system calibration procedure.
- **Drift** exists if properties of a sensor varies with time or other ambient factors. The term long time drift usually indicates slow property degradation of a sensor over a long period of time.
- **Linear Range** is defined as a range in which the non-linearity error of the sensitivity can be negligible. The linear range is relied on the specification of the system requirement.
- **Full scale range** is defined as range between the minimum and maximum output values of a sensor. Usually the output of a sensor reaches a limit if the input signal exceeds specified range.
- **Hysteresis** is an error caused by when the measured property reverses direction, but there is some finite lag in time for the sensor to respond, creating a different offset error in one direction than in the other.
- **Noise** is random fluctuation of the sensor output over measurement time.

2 Weak Signal Conditioning

The output voltage or the current signal of a sensor is usually small. Taking MEMS microphone (commonly used sensor for acoustic system) as an example, the maximum output voltage of the microphone is on the order of ~ 250 mV. Usually the system requires the signal dynamic range be larger than 90dB which means the minimum voltage to be processed is less than 10 μV. To process such a weak signal the following analog front end circuits are preferred as shown in Fig. 1. The signal from the sensor is firstly amplified and then digitized by the analog-to-digital converter (ADC). The pre-amplifier can greatly enhance the signal level of the output of the sensor and relaxes the design requirements of the ADC. However the design of pre-amplifier becomes the bottleneck. Several design issues must be taken into considerations.

- **Noise** may be the most important design issue to be considered in the front end circuits design. The output signal of the sensor is usually low frequency varied signal. At low frequency range, the $1/f$ noise of the circuit is significant especially in CMOS circuits. Although several techniques such as chopping stabilization (CS), correlated double sampling(CDS) and auto-zero(AZ) have already been proposed to effectively remove the low frequency $1/f$ noise, not all the cases are suitable to apply such techniques. Also taking signal conditioning circuits after MEMS microphone as an example, the output impedance of the sensor is quite large which means the driving ability of the sensor is poor. If any of the CS, CDS or AZ is applied, the fast switching network which is necessary in the above techniques will disturb the output of the sensor and even force the sensor into abnormal work conditions. As a result the only way to reduce the $1/f$ noise is by properly sizing the transistors in the amplifier. Large transistor size may result into large chip area and large parasitic which is harmful to the stability of the amplifier.
- **Power consumption** is another important factor in designing front end circuits. In many portable devices and biomedical devices, the energy resource is usually a small battery which is mandatory in order to reduce the volume and weigh of the devices. To extend the battery life, low power circuits design techniques are necessary. The power consumption of the pre-amplifier is determined by two factors. One factor is the noise. The required thermal noise level determines the minimum current of the input stage. Besides the input stage, the amplifier also has output stage to provide enough current sinking and sourcing ability to drive the sampling network of the analog-to-digital converter.
- **Driving and anti-alias noise folding** Sampling the output of the pre-amplifier may lead to noise folding which can raise the noise level in the baseband. To overcome the problem the pre-amplifier must be band limited. However limited bandwidth may slow down the speed of the amplifier and as a result, the settling performance of the amplifier will be degraded. To solve the problem, an RC anti-alias filter can be inserted between the pre-amplifier and the ADC. This filter can also serve as a low power buffer. The reason is that, when ADC starts to sample the output of the RC filter, the sampling capacitor is firstly connected to the capacitor of the filter and charge sharing happens before the pre-amplifier starts to settle. If this capacitor is large enough, the sampling capacitor can soon reach to an initial value near to the final value and the amplifier only needs to compensate the difference and so the current consumption requirement as well as the power consumption requirement can greatly be reduced.

3 Driving the World

With various sensors, the electrical systems can gather information from the ambient and then pass the information to its powerful digital signal processing engine. After processing, the systems should also react with the ambient. Some systems only need to give out digital signal and other devices can receive the signal and makes further processing step. However some other systems need to gives out analog signal in order to drive other devices which can converter the electric signals back into physical signals. For example, in acoustic systems, after signal processing, the system should drive the receiver which can convert the digital signal back into sound. The receiver usually has low input impedance in order that the energy received is large enough. To drive this low impedance receiver, the output stage of any electrical systems must have enough current sinking and sourcing ability. Class-A amplifier has good linearity but the standby current is quite large so as to provide enough transient current. Its power efficiency is low which is not suitable for many low power conditions. Class-AB amplifier can keep the standby current to a very small value. But the circuits design is relative complex and dedicate feedback loop must be designed to stabilize the quiescent current. Class-D amplifier has no standby current and is the most power efficiency output driver architecture. The output of such amplifier may be pulse-width-modulation (PWM) signal or pulse-density-modulation (PDM) signal. The receiver must have sufficient filtering ability so that high frequency noise introduced by sharp 0-1 or 1-0 transition in PWM or PDM signals can be suppressed.

Low Noise Low Power Amplifiers

Shupeng Zhong and Nianxiong Nick Tan

In the applications of high precision, most signal bandwidth is not very wide, several Kilo Hertz or even smaller. But the signal is sometimes too weak to be quantified by Analog-to-Digital converters (ADCs) directly due to the inherent noise of any ADCs, including KT/C noise, flicker noise, thermal noise, etc. So a low noise amplifier is needed before ADCs. As with most design, power dissipation is of great importance. To amplify very week signal with ultra low power is challenging. In this chapter, we will discuss how to design an ultra-low power low noise amplifiers for narrow-band applications.

1 Noise in a Chip

The main noises in a low noise amplifier are flicker noise and thermal noise. There are several sources that can contribute to the noise in a chip [1].

1.1 Flicker Noise

For narrow band signal application, the flicker noise is usually the most significant noise source in an amplifier. The noise power can be expressed roughly by the following equation.

$$V_n^2 = \frac{K}{C_{ox} WL} \frac{1}{f} \tag{1}$$

S. Zhong (✉)
Vango Technologies, Inc., Hangzhou, China
e-mail: zhongsp@vangotech.com

N. N. Tan
Institute of VLSI Design, Zhejiang University, Hangzhou, China

N. N. Tan et al. (eds.), *Ultra-Low Power Integrated Circuit Design*,
Analog Circuits and Signal Processing 85, DOI 10.1007/978-1-4419-9973-3_3,
© Springer Science+Business Media New York 2014

Where K is a process-dependent constant. W, L is the MOSFET's size, C_{ox} is the unity capacitance of gate oxide. So the flicker noise is inversely proportional to the frequency and MOSFET's area. In some low frequency application like 50 Hz signal, the flicker noise could be as high as several uV/sqrtHz with a nominal MOS size. This noise can be decreased by enlarging the area of MOSFET, but even 100 times larger area can only reduce the noise power by a factor of 10. Too big transistors will also increase the power dissipation besides increasing the die size.

1.2   Thermal Noise

Unlike flicker noise, the noise spectrum is flat, so it is also called white noise. The noise power density is much lower than flicker noise in low frequency. In CMOS integrated circuits, two kinds of thermal noise need to be considered, which are thermal noise come from physical resistors and thermal noise from MOSFETs.

Resistor Thermal Noise

The noise power can be expressed by

$$V_n^2 = 4kTR \tag{2}$$

where K is bolzman constant, T is the absolute temperature, and R is the resistance.

As it can be seen, the thermal noise power is proportional to temperature and resistor value. So the only way to decrease the noise is to lower the value of resistor.

MOS Thermal Noise

The noise power can be expressed by [2]

$$V_n^2 = \frac{4kT\gamma}{g_m} \tag{3}$$

Where K is bolzman constant, T is the absolute temperature, γ is a process related constant, and g_m is the MOSFET's transconductance.

As can be seen, the thermal noise is proportional to temperature and inversely proportional to MOSFET's transconductance. The transconductance could be expressed to be

$$g_m = \sqrt{2u_n C_{ox} \frac{W}{L} I_d} \tag{4}$$

Where I_d is the drain current of MOS device, So larger g_m means smaller thermal noise, also means more power consumption.

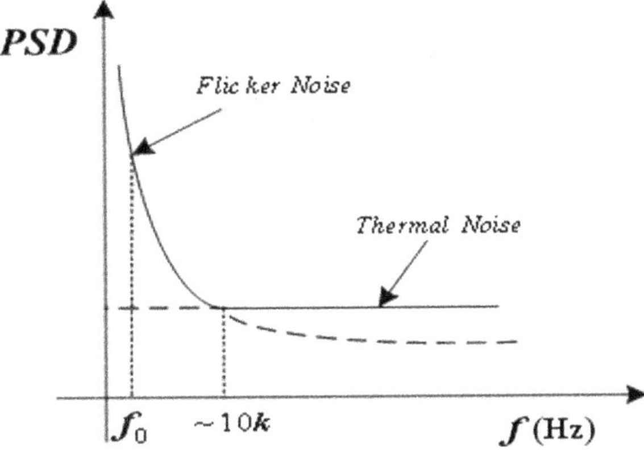

Fig. 1 Cross point of flicker noise and thermal noise

The combination of flicker noise and thermal noise in a typical MOS circuit is shown in Fig. 1. The cross point of flicker noise and thermal noise is normally locating at around 10 kHz, depending on the design and foundry process.

2 Offset and Low Frequency Noise Cancellation

There are several methods to decrease the noise power, especially the low frequency noise level, like Auto-Zeroing, Correlated Double Sampling (one type of Auto-Zeroing technique), Chopper Stabilization [3].

2.1 Auto-Zeroing

The basic principle of Auto-Zeroing is sampling the DC offset and low frequency noise, storing them onto some devices (usually capacitors), and then subtract them off from the noisy signal. Clock and capacitors are two necessary elements of auto zero circuits, so AZ structure is usually applied in switch-capacitor circuits [4].

There are several structures to realize the subtraction of DC offset and low frequency noise. One is subtracting the unwanted noise at the output of amplifiers, shown in Fig. 2.

In this structure, there are four switches named S1~S4. In the DC cancelling clock phase, all these switches are turn on, the offset of amplifier V_{os} is stored on the capacitor C0/C1 by an amount of $A_v V_{os}$, the outputs of AZ circuit named OUT1 and OUT2 remain zero, so in the top view, this amplifier does not have any offset.

But the above structure could be used in only low DC gain amplifier. In CMOS process, the DC offset is usually $1 \sim 10$ mV, determined by the size of MOS device and layout pattern. Assuming the DC gain of amplifier to be 60 dB, then the amplified offset which is stored on capacitors could be $1 \sim 10$ V, this voltage may be even

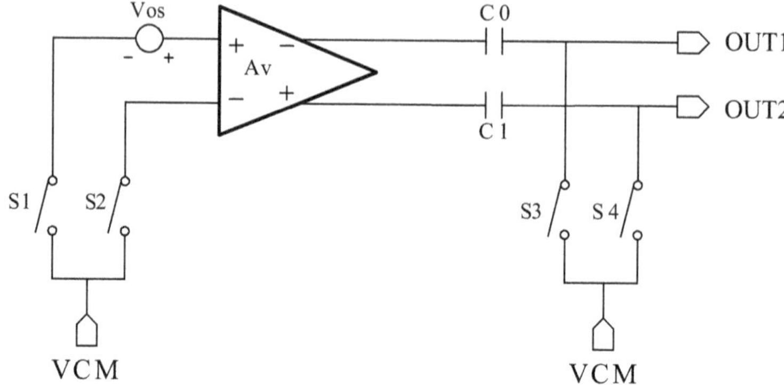

Fig. 2 Output-side DC cancellation

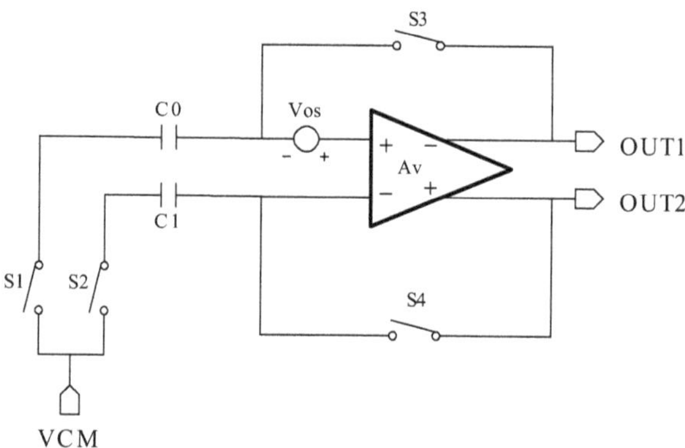

Fig. 3 Input-side DC cancellation

larger than source power. Lots of operational amplifiers have a DC gain higher than 60 dB for better performance, so another structure is needed for this high gain amplifier.

One way to solve this problem is moving the capacitors to the front side of amplifier, like the way shown in Fig. 3.

In the sampling phase, four switches are turned on, the voltage of V_{os} will be stored on C0 without amplification by a factor of A_v, avoiding saturation with large amplitude.

The two structures described above both have capacitors in their signal path, causing limited usage at both switch capacitor circuits and successive time circuits.

In order to avoiding the capacitor in signal path, an auxiliary amplifier is introduced in to make another cancellation path, and then put the capacitors in that path, shown in Fig. 4.

In this structure, the capacitors are placed at the input of auxiliary amplifier, the input voltage is V1, when

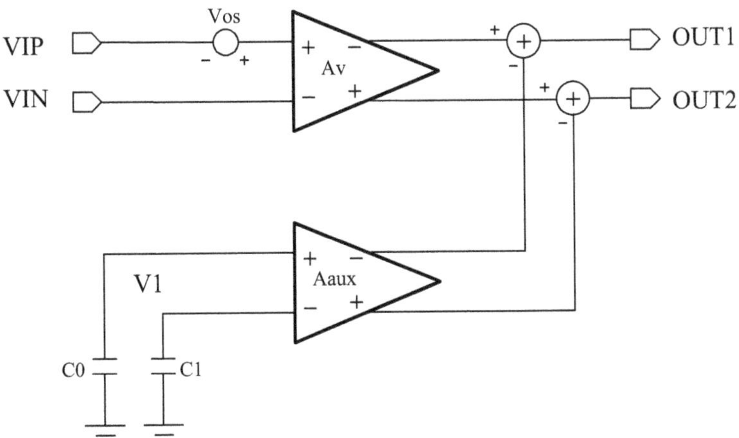

Fig. 4 DC cancellation with auxiliary amplifier

$$V1 = -\frac{A_v}{A_{aux}} V_{os} \tag{5}$$

The offset voltage of the main amplifier is cancelled by auxiliary amplifier perfectly.

2.2 *Chopper Stabilization*

Unlike Auto-Zeroing structure, chopper stabilization does not cancel the offset and low frequency noise by sampling and subtraction, but moving them or signal away from their original frequency to a higher place by modulation, and demodulating them to move back after finishing amplification. So the principle of chopper stabilization is separating the wanted signal and unwanted noise at spectrum level, then the separated noise can be filtered out by following low pass filter.

A clock is required in the chopper stabilization amplifier just like auto-zeroing, but capacitors are not needed, instead of resistors. This makes chopper stabilization more suitable in successive time application.

There are also several structures to realize chopper stabilization amplification. One common way is shown in Fig. 5 [5].

It's composed of eight switches, four resistors and one operational amplifier. The four switches at the front side of amplifier play a role of modulation of signal. The chopping clock has two phase named $\Phi1,\Phi2$, switches S1/S4 are controlled by phase $\Phi1$ and S2/S3 are turned on by $\Phi2$. The other four switches act as demodulation, S5/S8 are controlled by $\Phi1$ and S6/S7 turned on by $\Phi2$. Resistors together with operational amplifier amplify the modulated signal by a factor of R2/R1. The process of how chopper stabilization works in spectrum view could be shown as Fig. 6.

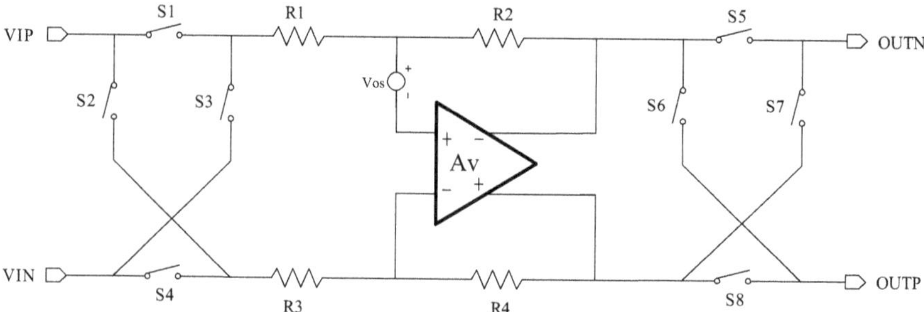

Fig. 5 Chopper-stabilization amplifier

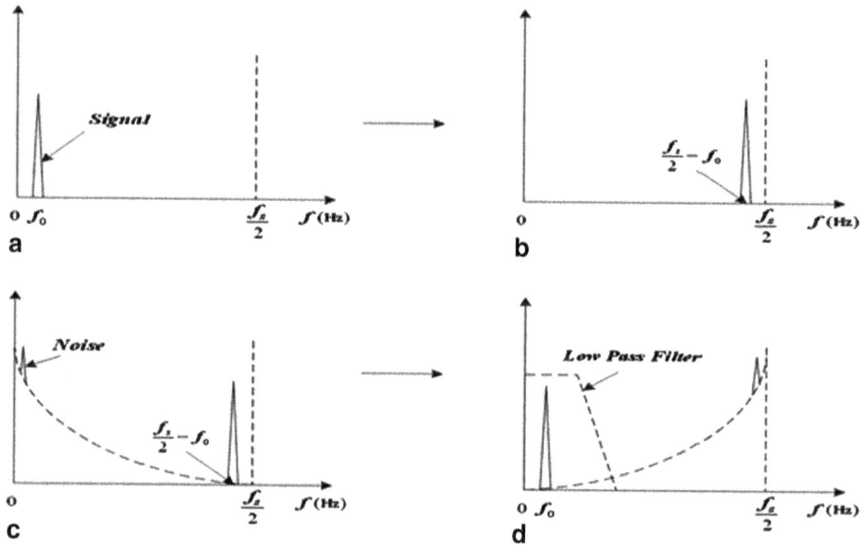

Fig. 6 Spectrum view of chopper-stabilization amplifier

Figure 6a shows the spectrum of input signal, it should be filtered by a low pass filter to wipe off high frequency noise or tones, or else they will be sampled down to low frequency by chopping clock. Figure 6b shows the modulated signal spectrum by the four switches at the input side. Assuming the chopping clock frequency to be f_c, then the signal band is moved to the place at $f_c - f_c$, where f_0 is the signal frequency. Usually f_c is set to be half of the ADC sampling frequency to avoid modulated tones sampled down to signal band, which is assumed to be f_s, so the modulated signal is located at $f_s/2 - f_0$. The modulated signal is then amplified by the closed loop amplifier by a factor of R2/R1, with all the noise source joined in, like resistor thermal noise, transistor thermal noise, and the most significant flicker noise, shown in Fig. 6c. But fortunately, the signal is placed far away from the low frequency noise. Then the noisy signal is demodulated by the same clock to move

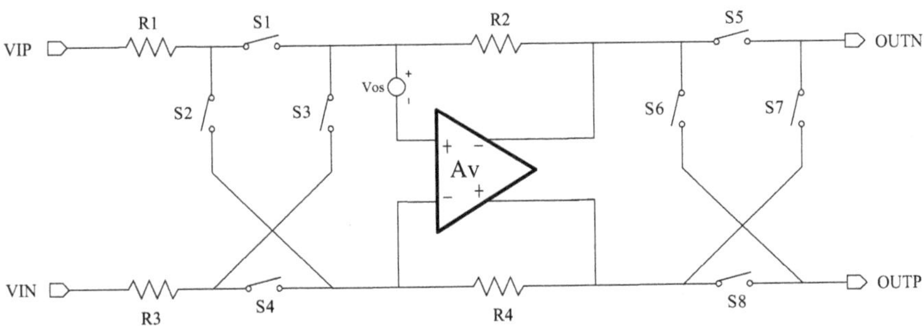

Fig. 7 Chopping at the input of operational amplifier

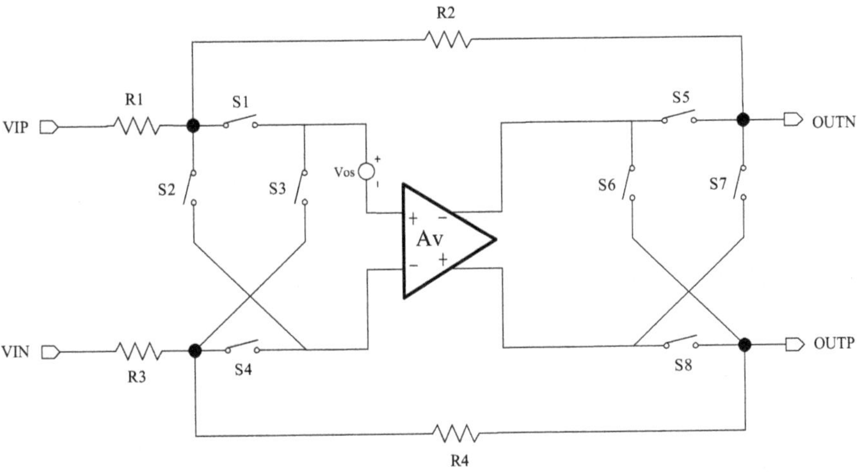

Fig. 8 Chopping inside the operational amplifier

back to its original frequency, the lower frequency noise is modulated to the place at $f_s/2$, far away from the signal frequency. Then it can be filtered out by a followed low pass filter, just like Fig. 6d shown. So the weak signal is amplified by chopping stabilization amplifier without mixing in the flicker noise, but the thermal noise at the frequency of $f_s/2 - f_0$ can not be removed away, it can only be decreased by circuits design, trading off between power consumption and device size.

There are two other methods to realize chopper stabilization, shown in Fig. 7 and Fig. 8.

In Fig. 7, the only difference between it and the first structure is that the modulating switches are placed at the right side of input resistors. So the current signal is modulated by chopping clock instead of voltage signal.

In Fig. 8, the noise and DC offset of operational amplifier are modulated by chopping clock instead of input signal. The chopping process at spectrum view is similar to the first structure introduced before.

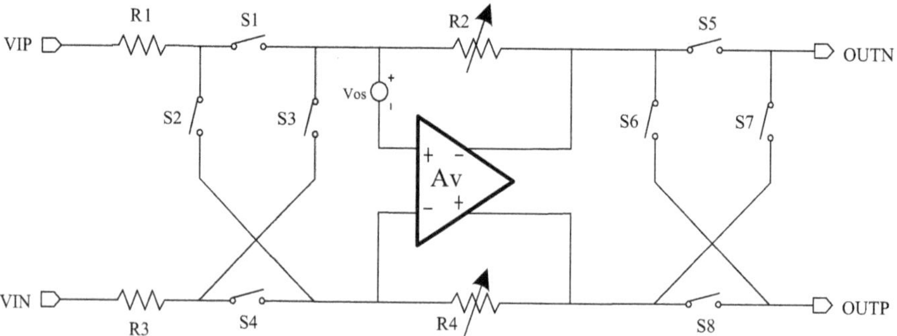

Fig. 9 Chopper-stabilization amplifier

3   Example

A low noise amplifier used in Energy-Metering IC is shown in this section. The input signal can be really weak, depending on the shunt resistance used in the energy meters. The frequency of signal is 50 Hz(or 60 Hz, varies in different countries), where the flicker noise dominates. In order to achieve enough accuracy, even with very small input, we still need enough signal to noise ratio. The required noise floor in this application should be less than -150 dBFS, assuming the reference voltage to be 1 V, then the noise needed to be less than 20 nV/sqrtHz, a lot smaller than the flicker noise at 50 Hz. The flicker noise of the low noise amplifier has to be reduced significantly in order to amplify the week input signal with enough accuracy.

In order to accomplish this goal, a low noise amplifier is designed shown in Fig. 9. It's almost the same with Fig. 7, the only difference is that resistors R2/R4 are designed to be programmable.

3.1   Resistors

The input resistor value of R1/R2 is set to be 5 k Ω. The thermal noise level of this resistor at room temperature is

$$V_{nR1} = \sqrt{4\,kTR} = \sqrt{4*1.38*10^{-23}*300*5\ k} = 9.1\ nV/sqrtHz$$

3.2   Switches

The switch type could be NMOS switch, PMOS or CMOS. The DC voltage at the switch site is the determined reason to choose which type to use. Although the

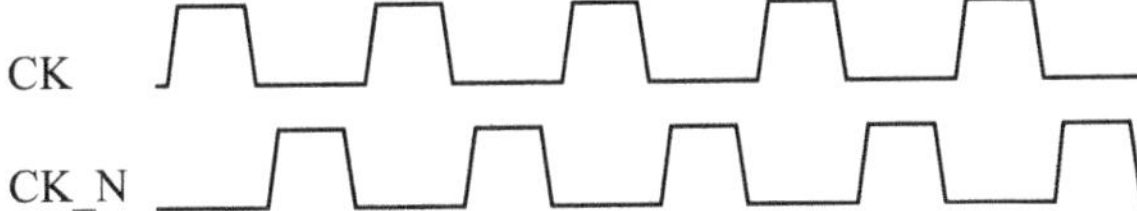

Fig. 10 Non-overlap clock

thermal noise of switch is not critical, but switches with too small size should be avoided.

3.3 Chopping Clock

The chopping clock is designed to be non-overlap style, shown in Fig. 10.

This type of clock is popular in switched-capacitor circuits such as ADCs. It can also be used in switch capacitor type common mode feedback in the operational amplifier.

3.4 Operational Amplifier

It's the most important module in chopper stabilization low noise amplifier. Lots of specifications of operational amplifier like DC gain, noise floor, signal bandwidth, will have impact on the final performance of low noise amplifier.

A two stage operational amplifier is designed for this low noise amplifier for the following reasons:

1. Achieve a better compromise between DC gain and signal swing, including output swing and input swing. In order to realize high gain, cascode structure must be applied to the amplifier, but in that structure, the output signal swing is not very large even the input signal's common mode voltage is set to be constant. Although one single stage with cascode structure can be designed to meet the specification, but the voltage margin of each MOS is not very sufficient, so in mass production, there will be some yield issues.
2. This operational amplifier needs to drive a resistor load. Because the input resistor is set to be 5 k Ω, assuming the closed-loop gain of low noise amplifier to be 16 times, then the output resistor is 80 k Ω, much smaller than the r_{ds} of a cascode amplifier.
3. There is an ADC following this low noise amplifier, the input capacitor of ADC could be several pF to narrow the signal bandwidth of amplifier if only one stage structure is applied.

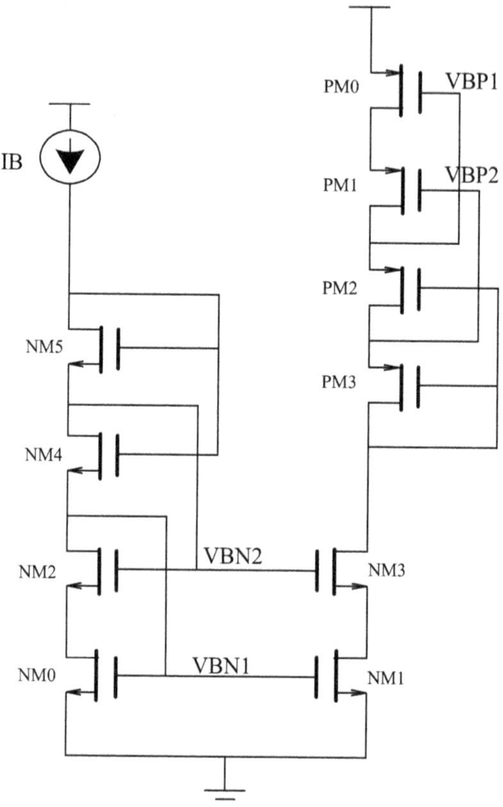

Fig. 11 Cascode bias generator

Current and Voltage Bias

The reference voltage and bias current are supplied by a Bandgap block. Inside this amplifier, we locally generate different bias current and bias voltage to make the amplifier work properly

The bias block structure is shown in Fig. 11. In order to achieve a high DC gain, we use a cascode structure in the first stage amplifier. The cascode amplifier needs two bias voltages for each PMOS and NMOS. Take NMOS for example, the one named VBN1 is the bias voltage for the bottom NMOS, performed as a current mirror. Another one called VBN2 is for the upper or cascoded NMOS. The goal of bias voltage is to bias these two cascode NMOS with wanted current and make them working in saturate state. Usually these two bias voltage need two bias path to generate, but in Fig. 11, this circuit generate them in one current path. A current from bandgap is fed into the drain of NM5, flow down through NM4/NM2/NM0, these 4 NMOS are connected in a serial way to generate bias voltage VBN1/VBN2. By selecting appropriate size of these four NMOS, all the NMOS could be biased to a saturate state.

The two bias voltages for PMOS named VBP1/VBP2 are generated by the same way.

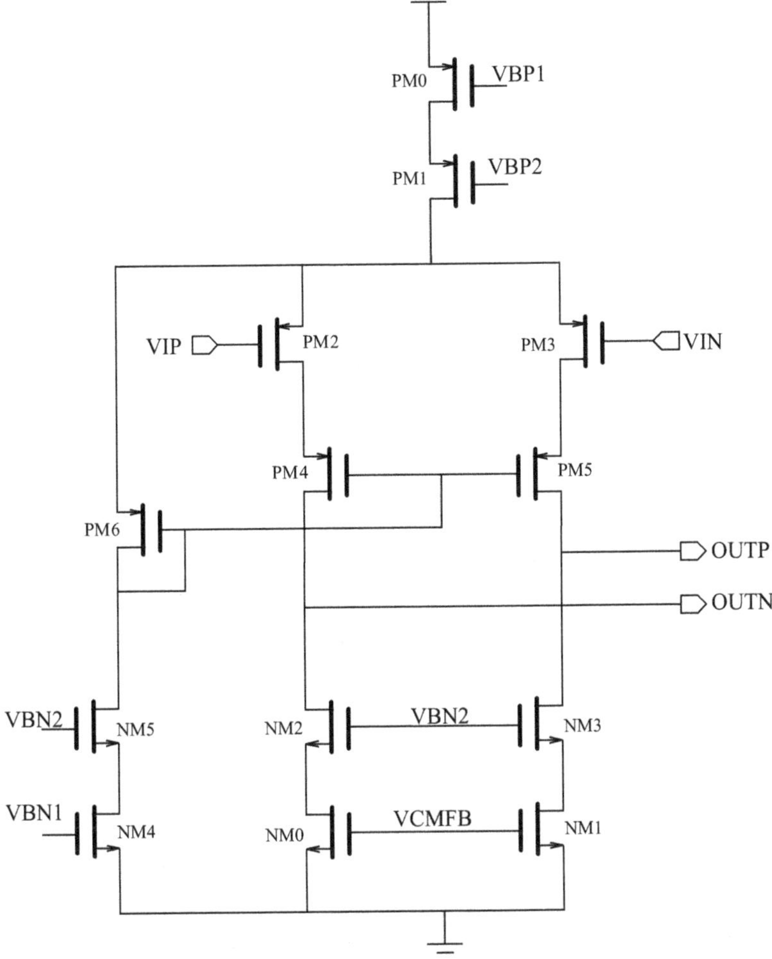

Fig. 12 First stage amplifer

First Stage

The first stage is designed to be cascode structure, shown in Fig. 12.

PM0/PM1 are the current source of this stage with cascode structure, the bias voltages of PM0/PM1 are supplied by VBP1/VBP2 generated from the bias circuit shown in Fig. 11.

The input MOS of this stage is P type for two reasons,

1. The flicker noise of PMOS is apparently lower than NMOS, so the MOS size could be designed to smaller than NMOS to achieve the same noise level.
2. This amplifier can work well even when the input signal's common mode voltage becomes zero volt which is the actual common mode voltage of shunt current sensor.

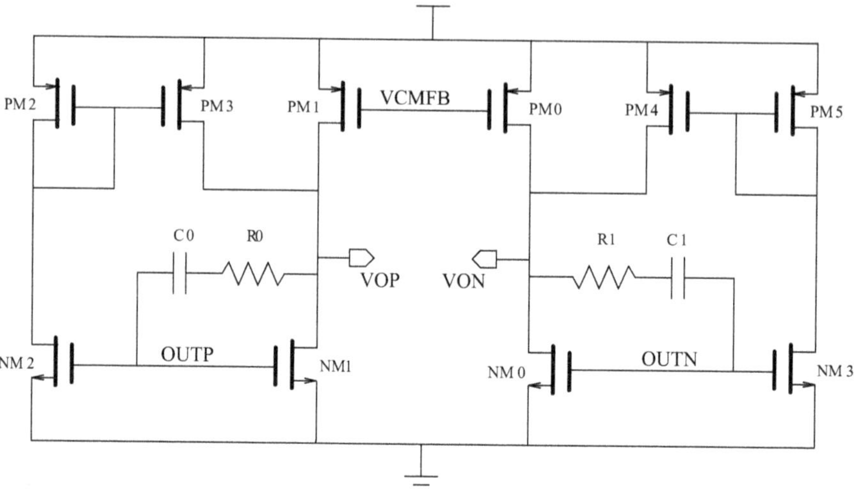

Fig. 13 Second stage amplifier

PM4/PM5 are bias by Vgs of PM6. The source node of PM6 is connected to the common source of input PMOS devices, so PM4/PM5 could be biased correctly even when the input common mode voltage of VIP/VIN varies.

The bias voltage of NM2/3 is also generated from the bias circuit shown in Fig. 11, named VBN2. NM0/1 are not biased by VBN1 because this amplifier is fully differential structure, NM0/1 must be biased by common mode feedback circuit, or else the amplifier will not work even a little bit mismatch exist between PMOS current mirror and NMOS current mirror.

Second Stage

As described before, the second stage should have large output signal swing and resistor driving capability, so a common source amplifier is a preferred structure. In order to save power dissipation, an AB class driver is used in this block. The final circuit of second stage is shown in Fig. 13 [6].

The gate voltages OUTP/OUTN are the output of the first stage, the common mode voltage of OUTP/OUTN is set to be quite low to bias NM0/NM1/NM2/NM3 to a state near weak inversion. So at the same current consumption, the size of NM0/NM1 could be much larger than regular structure. This could make the transconductance of NM0/NM1 larger, meaning faster slew rate and better phase margin. The sizes of NM2/NM3 are scaled down a few times than NM0/NM1 to save power dissipation.

PM3/PM4 are PMOS driving devices in this class AB output stage. They are biased by PM2/PM5. Because this stage is also fully differential, common mode feedback is needed just like stage 1. So PM0/PM1 are placed in parallel to PM3/PM4, acting as common current source. The size of PM0/PM1 could be designed to be small, as long as the current mirror mismatch could be compensated by these two PMOS current source.

Two stage amplifier has phase margin problem, so a miller capacitor is needed between the output and input of the second stage. By introducing this miller capacitor, the dominated pole is located at the output of first stage, with a approximately value of

$$p_1 = (r_{amp1} A_{amp2} C_0)^{-1} \tag{6}$$

Where r_{amp1} is the small signal output resistance of the first stage, A_{amp2} is the DC gain of second stage, C_0 is the value of miller capacitor.

The overall gain of this two stage amplifier is

$$A_v = g_{m1} r_{amp1} A_{amp2} \tag{7}$$

Where g_{m1} is the transconductance of the first stage. So the unity gain bandwidth is

$$GBW = A_v p_1 = \frac{g_{m1}}{C_0} \tag{8}$$

The second pole is located at the output of stage 2,

$$p_2 = \frac{g_{m2}}{C_L} \tag{9}$$

Where C_L is the second stage's capacitor load, g_{m2} is the transconductance of the second stage. Comparing to the GBW of this amplifier, the second pole is not far away from the unity gain frequency, even will be smaller than it when capacitor load C_L is big enough. So the phase margin could not be pretty in that case. Besides, the miller capacitor also brings in a positive zero located at

$$Z_0 = \frac{g_{m2}}{C_0} \tag{10}$$

This zero will also decrease the phase margin.

In order to obtain a better phase margin and larger unity gain bandwidth, a resistor is placed serial to the miller capacitor to introduce a negative zero to cancel the pole at the output of second stage. The negative zero is located at [7]

$$Z_1 = \frac{1}{(R_0 - g_{m2}^{-1})C_0} \tag{11}$$

Make $Z_1 = p_2$ the value of compensating resistor should be

$$R_0 = g_{m2}^{-1}(1 + \frac{C_L}{C_0}) \tag{12}$$

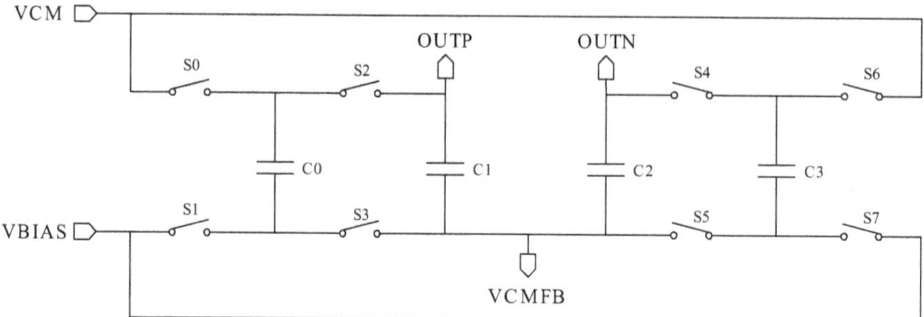

Fig. 14 Common mode feedback

Common Mode Feedback

Because the ADC followed this low noise amplifier is switch-capacitor type, the output signal of this amplifier is sampled by ADC in the specific clock time. So the common mode feedback circuit could be designed to be switch-capacitor type, Shown in Fig. 14.

It's composed of four capacitors and eight switches. The clock controlling these switches comes from non-overlap clock block, the same as the chopping clock. S0/S1/S6/S7 are controlled by phase $\Phi1$, S2/S3/S4/S5 controlled by phase $\Phi2$. By several clock cycles, the common mode voltage of OUTP/OUTN will follow to the voltage of VCM, and VCMFB will follow to VBIAS.

3.5   Specifications

The typical open loop gain of this two stage operational amplifier is about 120 dB, large enough to lower the distortion or enhance the PSRR. The GBW is about 25 MHz, so it can be sampled by a following ADC at 1 MHz with a closed loop gain of 16 times.

The noise including flicker noise and thermal noise at the chopping frequency is less than 10 nV/sqrtHz. So the overall noise considering operational amplifier and resistors is about 20 nV/sqrtHz.

The power consumption of this chopper stabilization amplifier is about 100 uA with the performance described above. If the chopping frequency becomes lower, or with a smaller closed loop gain, the power consumption could be designed much lower than 100 uA.

This module is used in a metering chip which has a whole system including ADC/PLL/MCU/DSP on it. It's produced in mass production at TSMC. So far, millions of this chip have been used in energy metering instruments and works quite well.

References

1. Behzad Ravazi; *"Design of Analog CMOS Integrated Circuits"*, McGraw-Hill Companies, Inc, New York, 2001. Chapter 7.
2. Mikael Gustavsson, J. Jacob Wikner and Nianxiong Nick Tan; *"CMOS Data Converters For Communications"*, Kluwer Academic Publishers, Chapter 6
3. Chiristian C. Enz, Gabor C. Temes; "Circuit Techniques for Reducing the Effects of Op-Amp Imperfections: Autozeroing, Correlated Double Sampling, and Chopper Stabilization", Proceedings of the IEEE, Vol. 84, No. 11, November 1996
4. Behzad Ravazi; *"Design of Analog CMOS Integrated Circuits"*, McGraw-Hill Companies, Inc, New York, 2001. Chapter 13.
5. R. Gregorian and G. C. Temes; *"Analog MOS Integrated Circuits"*, John Wiley & Sons, Inc., 1986.
6. H. A. Aslanzadeh, S. Mehrmanesh, M. B. Vahidfar, A. Q. Safarian, R. Lotfi; "High-Speed Low Power Pipelined A/D Converters using "Slew Boost" Technique", ISLPED 03, August 25–27, 2003, Seoul, Korea.
7. Paul R. Gray, Paul J. Hurst, Stephen H. Lewis, Robert G. Meyer; *"Analysis And Design of Analog Integrated Circuits"*, John Wiley & Sons, Inc. New York. Chapter 9.

Data Conversions

Liyuan Liu, Dongmei Li and Zhihua Wang

The $\Delta\Sigma$ type data converters are widely used in audio signal conditioning system and other low frequency signal acquisition systems in which large dynamic range signal conditioning is required. With the progress of CMOS technology, supply voltage continuously scales down with the feature size. The international technology roadmap for semiconductor (ITRS) shows that the operating supply voltage for CMOS transistors is 1-V currently and approaches to 0.6-V until 2020. The purpose for supply reduction is to prevent device from breaking down, because the gate oxide is becoming thinner and thinner. The $\Delta\Sigma$-ADC can convert analog signal into their high precision digital representations. Amplifier is the key building block and its noise as well as linearity limits the overall ADC performance. Low supply voltage results into small voltage headroom and swing for amplifier which may deteriorate dynamic range and cause unacceptable harmonic distortion. To overcome the above problems usually large power is needed. In this chapter, design of $\Delta\Sigma$-ADC under only 1-V supply is explored. Various techniques are adopted to achieve high resolution under low voltage while keeping the power dissipation under a low level. The function of a $\Delta\Sigma$-DAC is to reconstruct the analog signal and deliver enough power to the output load. Taking the digital hearing aid as an example, if large sound level is required, the peak output power can exceed tens of mill watts, which is an order larger than all circuits excluding the DAC. Hence more accurately speaking, low power DAC means that power delivery efficiency of the DAC is high. Adoption of class-D amplifier is a good way to achieve good efficiency. However fast switching operation may introduce extra noise and distortions, dedicated techniques should be employed.

L. Liu (✉)
Institute of Semiconductors, Chinese Academy of Sciences, Beijing, China
e-mail: liuly@semi.ac.cn

D. Li
Department of Electronic Engineering, Tsinghua University, Beijing, China

Z. Wang
Institute of Microelectronics, Tsinghua University, Beijing, China

N. N. Tan et al. (eds.), *Ultra-Low Power Integrated Circuit Design*,
Analog Circuits and Signal Processing 85, DOI 10.1007/978-1-4419-9973-3_4,
© Springer Science+Business Media New York 2014

1 Ultra-Low Power ADC Design Principles

For large dynamic range ADC, it is better to make the thermal noise dominate the SNR performance. The first integrator contributes most of the noise and distortion, and the amplifier inside is power hungry. In this part the power consumption of the first amplifier is derived and next low power design strategies are discussed. The derivations are based on [1].

The total thermal noise of the modulator (fully differential) can be expressed as:

$$N_T = \frac{2kT(1+2\gamma/3)}{OSR \cdot C_S} \tag{1}$$

In the equation, k is the Boltzmann constant (1.38×10^{-23} J/K), T is the room temperature (300 K), γ is the noise excessive factor of a specify amplifier and Cs is the input sampling capacitance of the first stage. If the reference voltage is $\pm Vr$, the maximum input signal amplitude is:

$$V_{i,\max} = OL \times V_r \tag{2}$$

The parameter OL is called overload level which indicates the stability of the modulator. Then the maximum signal to noise ratio can be written as:

$$SNR_{\max} = \frac{(OL \times V_r)^2 \times OSR \times C_S}{4kT(1+2\gamma/3)} \tag{3}$$

In the switched-capacitor integrator, the amplifier needs enough settling in half of the sampling clock period which requires the amplifier's unit gain bandwidth (GBW) be α times greater than sampling frequency:

$$GBW = \frac{g_m}{2\pi\beta C_L} \geq \alpha f_s \tag{4}$$

The factor g_m denotes the trans-conductance of the input amplifier and can be calculated as:

$$g_m = \frac{I_t}{V_{ov}} \tag{5}$$

It is the tail current for the input differential pair and V_{ov} is correspondence overdrive voltage. C_L is the load capacitance of the amplifier. If a two-stage amplifier with compensation is adopted, GBW is determined by compensation capacitor C_c which is usually a portion of C_L as indicating by factor β. The amplifier inside integrator is to transfer the charge on the sampling capacitor to the integration capacitor, the larger the sampling capacitance is, and the more currents are needed from the amplifier. So the sampling capacitor is the effective load capacitance to the amplifier during charge transferring phase.

The power consumption of the first amplifier can be expressed as (6). Power consumption for other braches beside tail currents are denoted by δ.

$$P = (1+\delta)I_t \times V_{DD} \tag{6}$$

Using (1) ~ (6), the following relation can be obtained:

$$P \geq \frac{16\pi kT\alpha\beta(1+\frac{2}{3}\gamma)(1+\delta) \times V_{ov} \times BW \times SNR_{max}}{OL^2 V_{DD}} \tag{7}$$

In the derivation, the reference voltages have been chosen as $\pm$VDD. For a certain applications, BW and SNR_{max} are specified and power dissipation of the first amplifier is impacted by the choice of following parameters:

- V_{ov} and V_{DD}. The power dissipation is proportion to the over drive voltage and inversely proportion to supply voltage. Commonly V_{ov} is chosen to be larger than 100 mV to prevent the transistors from entering sub-threshold range. In this case the lower the supply is, the larger the power will be which means low power design under low voltage is more difficult.
- OL. Increase the overload level of the modulator is helpful to low down the power consumption. Large OL means good stability of the modulator. So adopting multi-bit quantization is a good choice.
- α. This factor indicates GBW of the amplifier and determines settling error in half of the sampling period. So it is important to choose modulator whose performance is insensitive to the settling error in order to reduce the factor α. In this case single loop topology is a good candidate and amplifier's GBW as low as 1.5 times the sampling frequency is enough.
- β and δ. For single stage amplifier, GBW is determined by C_L, and $\beta=1$ and for two-stage amplifier the miller compensation capacitor C_c determines GBW. Usually C_c is half of C_L and $\beta=0.5$. Although tail current can be reduced by half in two-stage amplifier, the output branch current should be three times larger than tail current to ensure stability when amplifier is in feedback configuration. In the single stage amplifier, however, the output branch can have the same current as tail current. So to obtain more power efficient design it is more proper to choose single stage amplifier.

Although the single stage amplifier is preferred in low power design, its gain is relatively small and non-linear gain induced distortion and noise mixture effect should be taken into considerations. Since there is no headroom for gain increment, it is important to solve this problem on the architecture level. Feedforward topology with multi-bit quantizer hence becomes a good choice.

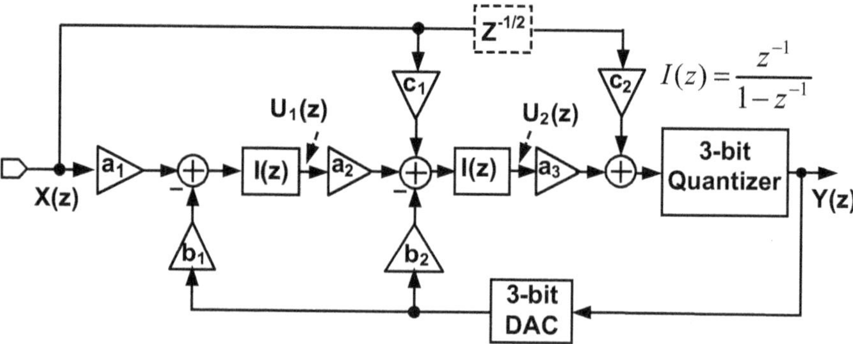

Fig. 1 Architecture of the $\Delta\Sigma$ modulator

2 Ultra-Low Power ADC Design Example

In this part we will give a low power $\Delta\Sigma$-ADC design example. The content of this part is based on [2]. In the following we show a $\Delta\Sigma$-ADC with 91.3 dB peak SNDR under 1-V supply with 360 μW total power dissipation. To overcome problems due to the limited voltage headroom, the CIFB topology with multi-bit quantizer is chosen. The advantage is the absence of signal component from integrators' outputs. Distortion due to non-linear amplifier gain is prevented however the quantization noise mixture should be taken into considerations. The noise transfer function (NTF) must be carefully chosen to mitigate such effect. Analog summing is necessary if there is a direct feed forward path. Traditional capacitive analog summing circuitry introduces attenuation and additional amplification or reference scaling is needed. In this design a single-capacitor summing circuitry is proposed. The effect owing to the parasitic capacitance is emphasized. In 2.1, the modulator architecture is presented. The choice of NTF is discussed. In section 2.2, detailed circuits design is described. In 2.3 the implementation of the decimation filter is explained. In 2.4 experiment results are shown.

2.1 *Architecture of the Modulator*

The architecture design of the $\Delta\Sigma$ modulator is presented in this section. Low power considerations are explained firstly followed by a detailed description of the proposed modulator. Non-linear gain is analyzed in the next.

Low Power $\Delta\Sigma$ Modulator

The CIFB modulator with direct input feed forward path [3] is adopted as shown in Fig. 1. The advantage of feed forward path is the absence of signal component at integrators outputs left the integrators only processing the quantization error [4].

The voltage swings of the integrators are hence reduced. By further employing a 3-bit quantizer, the quantization step decreases and swings get even smaller. A half-period delay exists in the feed forward path which relaxes the timing requirement of the quantizer [5]. Such a delay does not alter the characteristic of the modulator much which will be described later. Reduction of swings relaxes gain requirement of the amplifier. Simple amplifier such as single stage amplifier [6, 7] or even inverter based amplifier [8] can be used to construct the integrator and hence power is saved. Without this feature, the cascaded technique has to be adopted to increase amplifier gain under low voltage. More power is needed for two reasons. Firstly, more branches exist between the power rail and ground. Secondly large current is required to push additional poles associated with the cascaded stages far away to ensure amplifier stability in feedback configuration.

Coefficients of the Modulator

By replacing the quantizer with its linear model, signal transfer function (STF) and NTF can be expressed as

$$STF(z) = \frac{c_2(z-1)^2 z^{-1/2} + a_3 c_1(z-1) + a_1 a_2 a_3}{(z-1)^2 + a_3 b_2(z-1) + a_2 a_3 b_1}, \tag{8}$$

And

$$NTF(z) = \frac{(z-1)^2}{(z-1)^2 + a_3 b_2(z-1) + a_2 a_3 b_1} \tag{9}$$

Coefficients a_1 a_2 b_1 b_2 and c_1 are design parameters and are determined by capacitance ratios. Coefficients c_2 and a_3 should be one ideally, however parasitic may alter them. Changes of these coefficients would modify the modulator characteristic which should be considered. If a_1 and c_1 are equal to b_1 and b_2 respectively the STF has a flat unity gain over dc to half of the sampling frequency (f_s) without the half-period delay. Considering the delay, the STF will be different. Because of oversampling, signal range is from dc to $0.5 f_s/\text{OSR}$ where OSR represents the oversampling ratio. In this design, OSR is chosen as 128. As a result, the signal range is close to dc and is very narrow. Hence both $(z-1)^2$ and $(z-1)$ have strong attenuation and relations in (10) are satisfied.

$$\left| c_2(z-1)^2 z^{-1/2} \right| \ll \left| a_3 c_1(z-1) \right| \ll a_1 a_2 a_3$$
$$\left| (z-1)^2 \right| \ll \left| a_3 b_2(z-1) \right| \ll b_1 a_2 a_3 \tag{10}$$

In this case, according to (8) the STF in the signal range still has an almost flat unity gain response. Obviously, according (9), such a delay does not have any influence on the NTF.

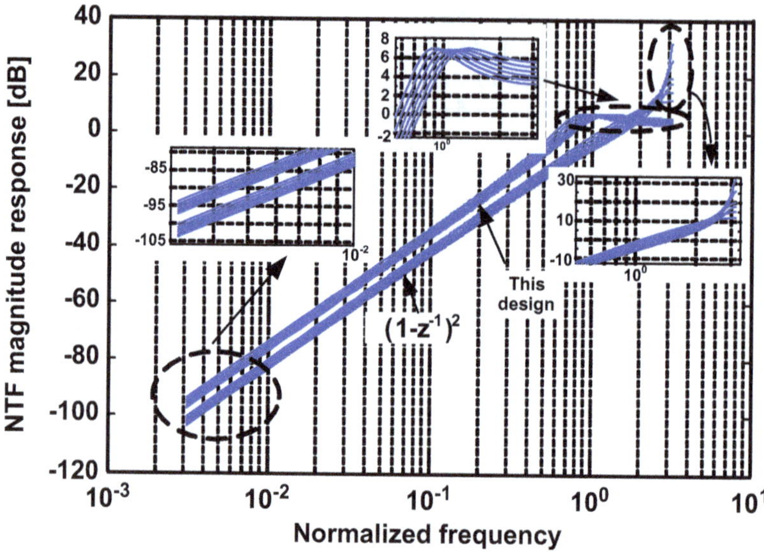

Fig. 2 Magnitude response of two kinds of NTFs with variations of a_3

The z-domain integrators' outputs can also be derived as

$$U_1(z) = \frac{a_1(1-c_2 z^{-1/2})(z-1)X(z)}{(z-1)^2 + a_3 b_2(z-1) + a_2 a_3 b_1} \\ - a_1 I(z) NTF(z) Q(z)$$

(11)

and

$$U_2(z) = \frac{\left[c_1(z-1) + a_1 a_2\right](1-c_2 z^{-1/2})X(z)}{(z-1)^2 + a_3 b_2(z-1) + a_2 a_3 b_1} \\ + \left[\frac{NTF(z)-1}{a_3}\right]Q(z)$$

(12)

In (11) and (12) the signal component exists. If c_2 is strictly equal to one, the signal component $X(z)$ in $U_1(z)$ and $U_2(z)$ is shaped by $(1 - z^{-1/2})(z - 1)$ and $(1 - z^{-1/2})$ respectively and is hence negligible. If c_2 is altered, e.g. by parasitic, $X(z)$ in $U_1(z)$ is still attenuated by $(z-1)$ but there is significant $X(z)$ leakage in $U_2(z)$ owing to non-zero c_2. As a result, the output swing of the second integrator increases. In III-B a technique will be described to keep c_2 equal to one regardless of the parasitic.

With $a_1 = b_1 = 1$, $c_1 = b_2 = 2$ and $a_2 = a_3 = 1$, the modulator has an NTF of $(1 - z^{-1})^2$. The maximum pass band gain is as high as 12 dB and the noise shaping ability of such NTF is aggressive. However if non-idealities cause a_3 to vary, e.g. from 1.0 to 1.5, the maximum pass band gain varies from 12 dB to over 30 dB as shown in Fig. 2. The problem is that non-linear amplifier gain causes severe noise mixture. In

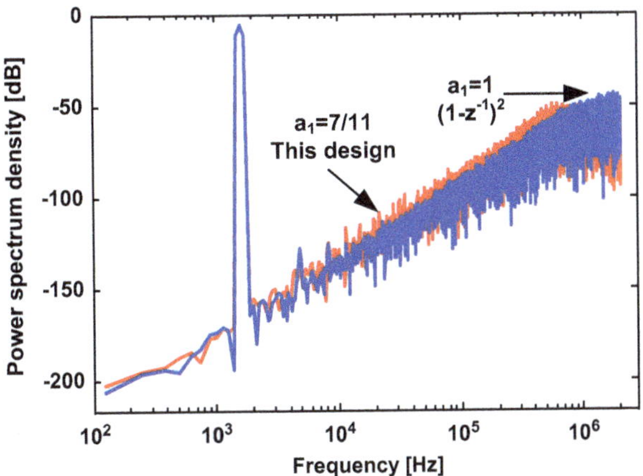

Fig. 3 Ideal output spectrum with different NTFs

this work, $a_1 = b_1 = 7/11$, $c_1 = b_2 = 7/8$ and $a_2 = 6/8$ are chosen. As depicted in Fig. 2, unlike $(1-z^{-1})^2$ with its maximum pass band gain at $f_s/2$, the proposed NTF has a ripple in the pass band. The gain at the peak of the ripple is 7 dB and it drops to 3.5 dB at $f_s/2$. The advantage is that NTF changes slightly with variation of a_3. As a result, modulator with such NTF has more immunity to the non-linear gain effect which will be described later.

Ideal output spectrum of the modulator with the proposed NTF is depicted in Fig. 3. With OSR of 128, simulated maximum signal to quantization noise ratio (SQNR) with 1.8 Vpp input at 1.625 kHz is 103.7 dB. The peak SQNR with NTF $=(1-z^{-1})^2$ under the same input is 107.5 dB which is also plotted in the figure.

Non-Linear Amplifier Gain

The modeling of non-linear gain effect can be found in literatures such as [9–13]. The transfer function of an integrator considering finite amplifier gain is show in (13) [9].

$$\frac{U(z)}{V(z)} = \frac{(2\dfrac{A-1}{A}-1)z^{-1}}{1-\dfrac{A-1}{A}z^{-1}} \tag{13}$$

In the above expression, $V(z)$ and $U(z)$ represent z-domain input and output of the integrator respectively. Finite amplifier gain is denoted as A. The time domain relation can be derived as (14) which shows that there is an error component due to A.

$$u(n) = \underbrace{u(n-1) + v(n-1)}_{\text{Ideal component}}$$

$$\underbrace{-\frac{1}{A}\left[2v(n-1) + u(n-1)\right]}_{\text{Error component}} \tag{14}$$

In fully differential implementation, non-linear amplifier gain can be modeled as a fourth-order polynomial with even order coefficients [13] as expressed in (15).

$$A = A_0\left[1 - \varepsilon_2 u^2(n) - \varepsilon_4 u^4(n)\right] \tag{15}$$

A_0 is the dc gain and ε_2 and ε_4 are the second and fourth order coefficients.

The non-linear gain plays a role as a mixture. With feed forward path, integrators only process quantization error signal. The mixture will cause high frequency quantization noise to fold back into the baseband and increase the noise power. Using (13) ~ (15), the baseband noise power is simulated. Figure 4a shows the simulation results with NTF of $(1 - z^{-1})^2$ and $a_3 = 1$, the baseband noise changes significantly if the gain is less than 50 dB. The same phenomenon is observed in [5]. Every 10 dB increment of the dc gain leads to about 10 dB reduction of the noise power. In Fig. 4b the variation of a_3 is considered. In circuits' implementation, ε_2 and ε_4 can be as large as 0.5 and 1.0 respectively and a_3 can approach 1.3. As a result, larger than 60 dB dc gain is required to make the noise power lower than -105 dB. With the proposed NTF however, variations of the excessive noise due to ε_2 and ε_4 as well as A_0 are much smoother. DC gain of 40 dB is enough to keep the noise power lower than -114 dB even if a3 is 1.5 as depicted in Fig. 4c, d.

Employing high gain amplifier can thoroughly eliminate the non-linear gain effect. Under low voltage, it is difficult to achieve high gain and low power simultaneously. Applying modulator with feed forward path combined with multi-bit quantizer is an effective way, however, the NTF should be properly chosen. Otherwise, the requirement of amplifier gain is still stringent.

2.2 Circuits Design

The switched-capacitor representation of the modulator is depicted in Fig. 5. The stray insensitive inverting integrator is utilized, in which switches are controlled by a double phase non-overlapping clock. The total sampling capacitance (C_S) is around 2 pF for low thermal noise. Table 1 lists all the capacitance values in the modulator. Each integrator samples at Φ_1 and transfers charges at Φ_2. For 16 kHz bandwidth and OSR of 128, f_S is chosen as 4.096 MHz. The first amplifier is chopped at 128 kHz ($f_S/32$) to remove the flicker noise. Chopping happens in the middle of the sampling phase of the first integrator in order to prevent interferences [14].

Output of the second integrator is quantized into 3-bit digital codes by a flash quantizer composing of seven comparators. The reference voltage for each compa-

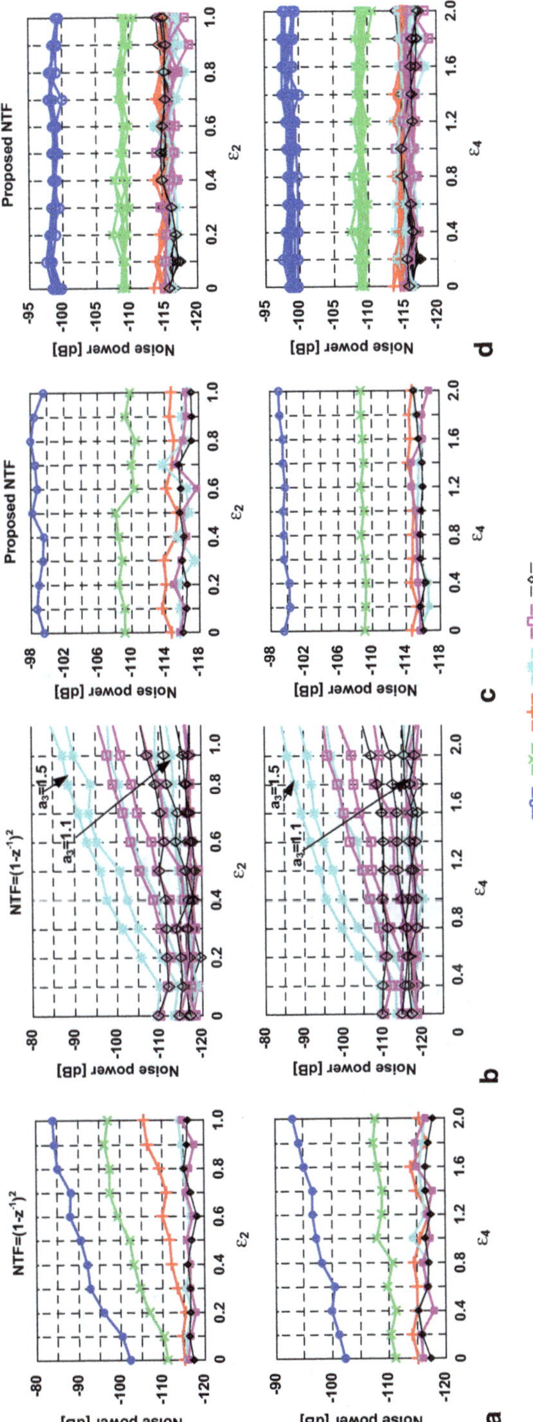

Fig. 4 Simulated baseband noise power owing to finite dc gain and non-linear gain coefficients ε_2 and ε_4 with **a** NTF$=(1-z^{-1})^2$ $a_3=1$. **b** NTF$=(1-z^{-1})^2$ a_3 varies from 1.1 to 1.5. **c** proposed NTF $a_3=1$. **d** proposed NTF a_3 varies from 1.1 to 1.5

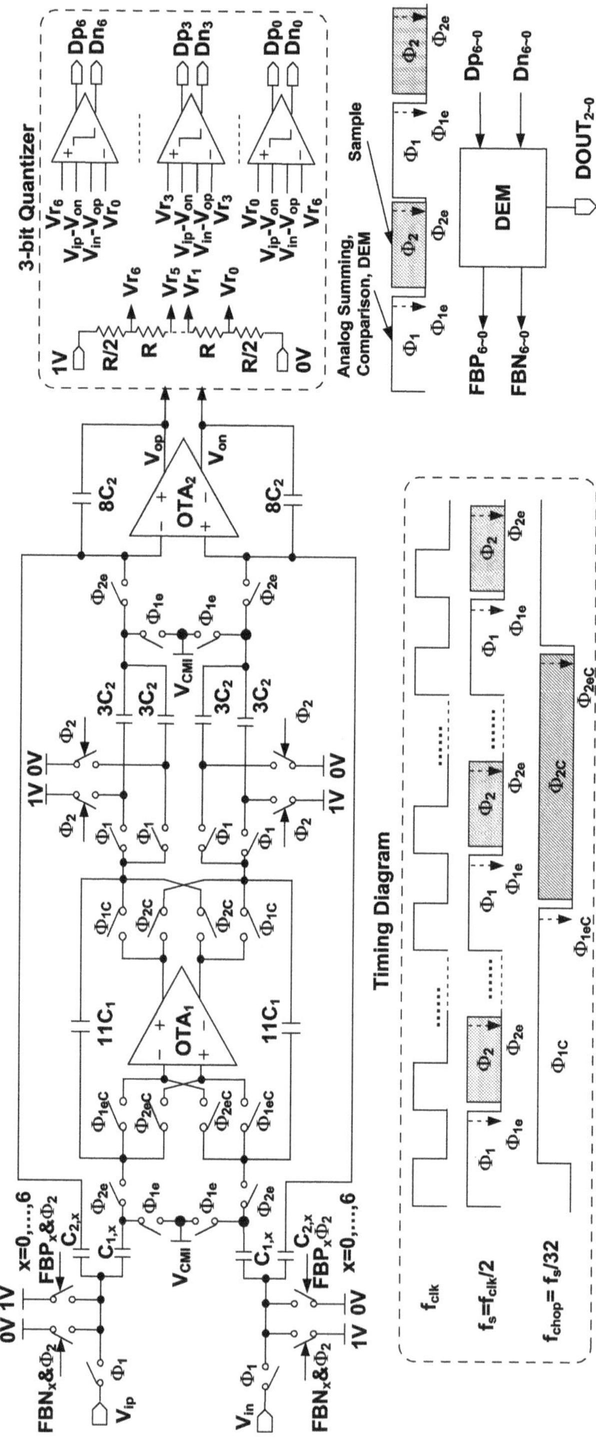

Fig. 5 Switched-capacitor circuits of the modulator

Table 1 Capacitance values in the modulator

Coefficients	Capacitance ratios	Capacitance values
a_1, b_1	$\sum_{x=0}^{6} C_{1,x} / 11C_1$	C1 = C1, x = 260 fF
a_2	$6C_2 / 8C_2$	C2 = 75.6 fF
c_1, b_2	$\sum_{x=0}^{6} C_{2,x} / 8C_2$	C2, x = 75.6 fF

rator is provided by a resistor ladder. The digital codes are feedback into the loop through a 3-bit DAC which is implemented by splitting the sampling capacitor into seven identical units. The dynamic element matching (DEM) is employed to eliminate the mismatch among all the sub capacitors. The well known data weighted averaging (DWA) algorithm [15] is adopted.

Usually, the quantizer should sample the last integrator's output at Φ_1, finish analog summing and make the decision at Φ_2. The DEM algorithm should also be finished at Φ_2. Such timing requires high speed comparator and fast DEM. By introducing the half-period delay, the quantizer can sample at Φ_2 and quantization as well as DEM operation can occupy the entire Φ_1. This slight change of timing relaxes the speed requirement on the comparator and DEM circuit.

The Operational Trans-Conductance Amplifier

Because of the swing reduction, a high dc gain amplifier is thus not necessary. Current mirror amplifier shown in Fig. 6a can be used to construct the integrator. In such an amplifier, the first stage output has low impedance which contributes the non-dominant pole. The second stage which connects directly to the large load capacitance generates the dominant pole. Thus, the amplifier functions just like a single stage amplifier. Its stability in feedback configuration is guaranteed by the load compensation. Each output branch consumes the same current (I_o) as the tail current (I_T). If a two-stage amplifier is required to increase the amplifier gain, to make the amplifier stable in the feedback loop, the miller compensation is necessary. In order to get sufficient phase margin, the non-dominant pole should be at least three times larger than the gain bandwidth product (GBW) and the current for the output branch should be three times larger than the tail current. Compared to a current mirror amplifier, actually a large portion of current is wasted.

In the implementation, transistors M_{ca} and M_{cb} are added to form a positive feedback and provide some increment for the dc gain and GBW performance as shown in (16) and (17) [6], where g_{m1} is the trans-conductance of the input transistor, R_o is the output impedance and C_L is the load capacitance.

$$A_0 = g_{m1} R_o \frac{W_3 / W_2}{1 - W_c / W_2} \tag{16}$$

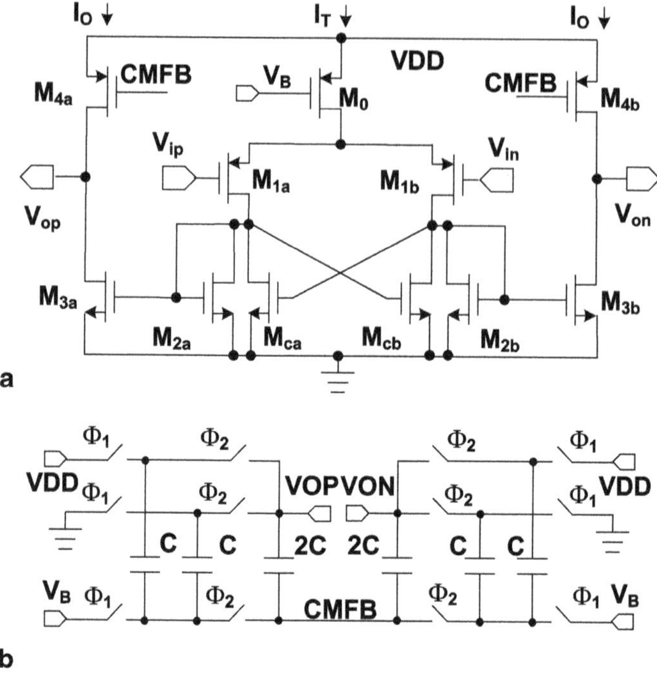

Fig. 6 a Amplifier core circuits. **b** Switched-capacitor common mode feedback circuits

Table 2 Transistor size of the first stage amplifier (W/L)

M0	M1a, 1b	M2a, 2b	M3a, 3b	M4a, 4b	Mca, cb
16/0.8	14.4/1.2	4.8/1.5	12/1.5	16/0.8	1.2/1.5

$$GBW = \frac{g_{m1}}{2\pi C_L} \frac{W_3 / W_2}{1 - W_c / W_2} \tag{17}$$

In this design all the nMOS transistors have the same channel length L of 1.5 μm as listed in Table 2. Since high gain amplifier is not required by this design, W_3/W_2 is chosen to be 10/4 and W_c/W_2 is set to 1/4 which provides only 10.4 dB improvement for the dc gain and 3.3 times enlargement for the GBW. In [6], W_c/W_2 is selected as 0.8 and the gain enhancement is as high as 30 dB. The designed amplifier has a dc gain of 55 dB. The problem is that phase margin of the amplifier is only 29° even with a load capacitance of 3 pF. There is some signal ringing when the load gets smaller. As a result a potential instability exists.

The total quiescent current of our amplifier is 30 μA. The tail current is 10 μA and two output branches share the remaining 20 μA. All the transistors have an overdrive voltage of merely 100 mV. The input common mode voltage (VCMI) is set to 150 mV which is generated by a resistor ladder between supply and ground. The simulated dc gain of the amplifier is 39.8 dB. When the amplifier drives a load capacitance of 2 pF, the GBW is 15 MHz and the phase margin is 72°.

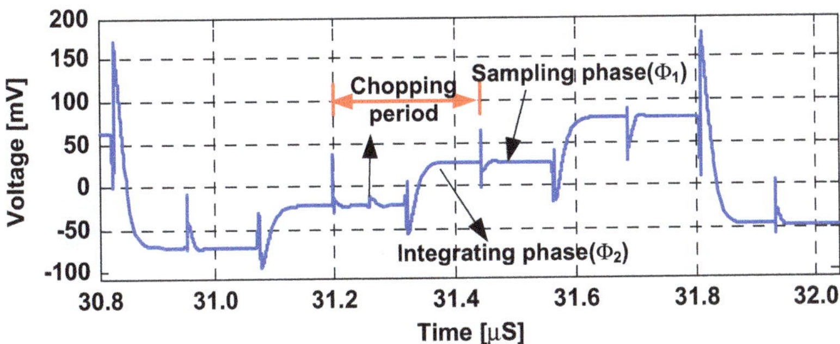

Fig. 7 Settling behavior of the first integrator

The output common mode voltage is stabilized using capacitive common mode feedback (CMFB) circuitry [16]. Unlike the conventional one, the charge sharing CMFB is used as shown in Fig. 6b. During Φ_1, one capacitor is charged to V_{DD} while the other is discharged to *GND*. During Φ_2, common mode voltage of $V_{DD}/2$ is generated by charge sharing.

Figure 7 shows the transient simulation of the settling behavior of the first integrator. The amplifier can settle down in half of the sampling period. The settling time is shorter in the sampling phase because of smaller load capacitance. When chopping happens, the amplifier output has a little disturbance, but the voltage can restore immediately.

One important design consideration is the noise contributed by the amplifier. The total noise NT from the first stage integrator can be calculated as (18) [1].

$$N_T = \frac{2kT(1 + 2\gamma\delta/3)}{OSR \cdot C_S} \tag{18}$$

In the above expression, k is the Boltzmann constant (1.38×10^{-23} J/K) and T is the absolute temperature (300 K). N_T is composed of two parts. One is the thermal noise contributed by switches and the other by the amplifier. Factor γ is thermal noise excessive factor of the current mirror amplifier. It can be found by noise analysis. In this design it is calculated to be 2.64.

Although most flicker noise is removed by chopping, there is still some residue noise left in the signal band and this effect is modeled by the factor δ. According to the analysis in [17] δ can be expressed as

$$\delta = (1 + 0.8525 f_k T_C) \tag{19}$$

In the above equation f_k is the corner frequency of the amplifier and T_C is the chopping period. Noise simulation shows that f_k is approximately 100 kHz and δ can be calculated to be 1.66. The total noise N_T can be kept below -99 dBV.

DC sweep analysis of the amplifier is performed to find out the amplifier gain dependence on the output voltage as illustrated in Fig. 8. Mismatch is considered.

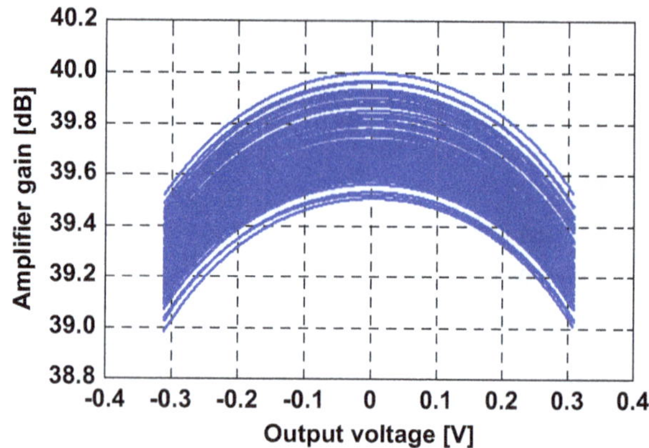

Fig. 8 Non-linear gain with the output voltage

Table 3 Extracted dc gain and even order coefficients

	Dc gain	ε_2	ε_4
Min	94.5	0.43	1.15
Max	100	0.47	1.43
Average	97.0	0.45	1.23

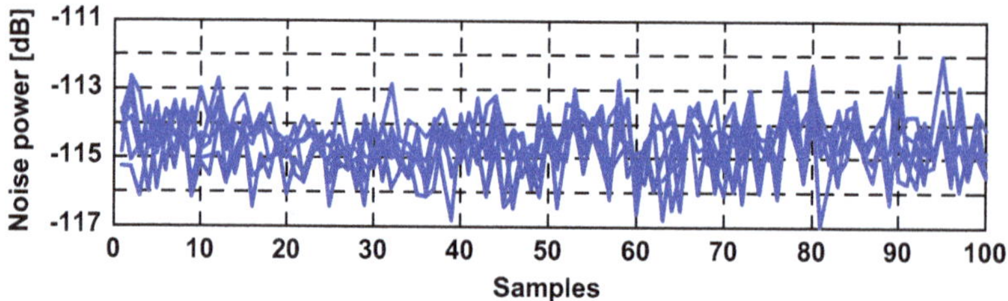

Fig. 9 Simulated noise power with non-linear gain

Coefficients (ε_2 and ε_4) within the output range from -0.3 to 0.3 V are extracted by the polynomial fitting method as summarized in Table 3. Substituting those coefficients into the behavior model, simulation results indicate that the baseband noise power is smaller than -112 dB as shown in Fig. 9.

Quantizer

The analog summing circuitry is necessary ahead of the quantizer to add direct input signal with output of the last integrator. Capacitive analog summing based on charge sharing [7, 8, 18] is popular because of its simplicity and low power consumption. However charge sharing causes signal attenuation which calls for additional amplification [5] or reference scaling [19]. Another method is digital summing [20]. In such a method, the last integrator's output and direct input feed forward signal

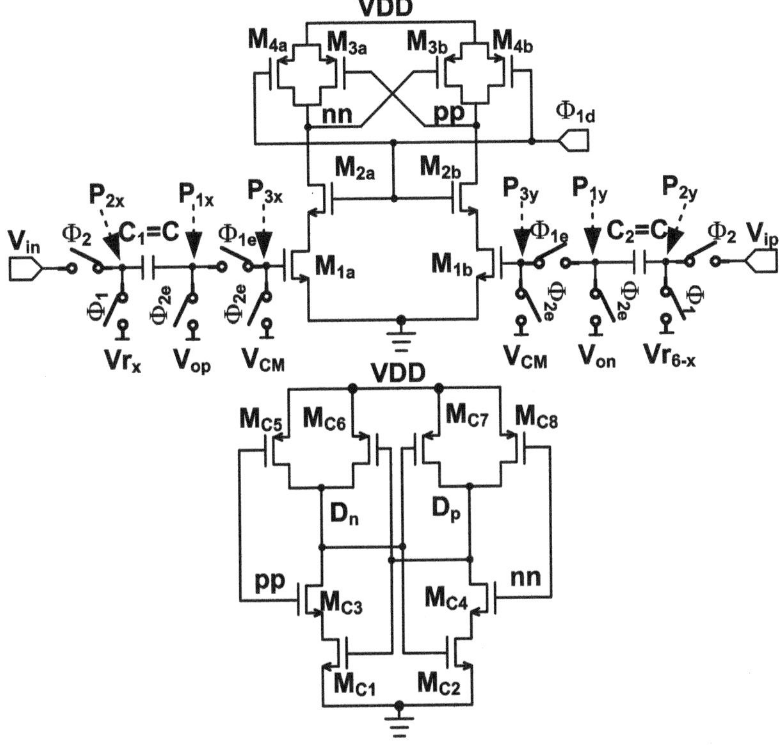

Fig. 10 Comparator with single-capacitor analog summing

are firstly digitized by separate quantizers and summation takes place in the digital domain. The main disadvantage is that additional quantizer placed along the input signal path brings extra quantization noise. To overcome such problem, the noise cancellation technique is needed.

In this work, the single-capacitor analog summing is adopted [21] as show in Fig. 10. The comparator in the figure is dynamic latch type [6, 7].During Φ_2, $V_{op}(V_{on})$ and $V_{in}(V_{ip})$ are connected to $P_{1x}(P_{1y})$ and $P_{2x}(P_{2y})$ respectively. At the same time, $P_{3x}(P_{3y})$ which is the input of the comparator is connected to the common mode voltage (VCM). The parasitic capacitors at $P_{1x}(P_{1y})$ and $P_{3x}(P_{3y})$ are also charged. During Φ_1, $V_{rx}(V_{r6-x})$ is connected to $P_{2x}(P_{2y})$ which forces the voltage at $P_{3x}(P_{3y})$ settle to a final value. Because of low power requirement, unit resistor in the reference voltages generation circuit is large which leads to large output impedance at $V_{rx}(V_{r6-x})$. As a result, owing to clock feed through, there are some glitches at $P_{3x}(P_{3y})$. The comparison result will be wrong, if the comparator starts to work before the voltage at $P_{3x}(P_{3y})$ settles. To prevent this problem, a deliberate delay signal Φ_{1d} instead of Φ_1 is used to activate the comparator. A latch is placed after the comparator. The decision result is stored and feedback during the next Φ_2.

By applying the charge conservation law, after settling, voltages at P_{3x} and P_{3y} are

$$V_{P_{3x}} = \frac{C + C_{P1}}{C + C_{P1} + C_{P3}} V_{op} - \frac{C}{C + C_{P1} + C_{P3}} V_{in}$$
$$+ \frac{C}{C + C_{P1} + C_{P3}} Vr_x + \frac{C_{P3}}{C + C_{P1} + C_{P3}} V_{CM}, \tag{20}$$

and

$$V_{P_{3y}} = \frac{C + C_{P1}}{C + C_{P1} + C_{P3}} V_{on} - \frac{C}{C + C_{P1} + C_{P3}} V_{ip}$$
$$+ \frac{C}{C + C_{P1} + C_{P3}} Vr_{6-x} + \frac{C_{P3}}{C + C_{P1} + C_{P3}} V_{CM}. \tag{21}$$

The comparator output can then be derived as

$$\begin{aligned}
D_p &= sign(V_{P_{3x}} - V_{P_{3y}}) \\
&= sign\Big[\frac{C + C_{P1}}{C + C_{P1} + C_{P3}} (V_{op} - V_{on}) \\
&\quad + \frac{C}{C + C_{P1} + C_{P3}} (V_{ip} - V_{in}) \\
&\quad - \frac{C}{C + C_{P1} + C_{P3}} (Vr_x - Vr_{6-x}) \Big] \\
&= sign\Big[\frac{C + C_{P1}}{C +} (V_{op} - V_{on}) \\
&\quad + (V_{ip} - V_{in}) - (Vr_x - Vr_{6-x}) \Big].
\end{aligned} \tag{22}$$

In (20) ~ (22), the value of C is 103 fF. C_{P1} and C_{P3} are parasitic capacitance from $P_{1x}(P_{1y})$ and $P_{3x}(P_{3y})$ to ground which are contributed by interconnected wires and MOS switches connected to those nets.

Because direct input signal and the reference voltage are both attenuated by the factor $C/(C + C_{P1} + C_{P3})$, there is an equivalent gain of $(C + C_{P1} + C_{P3})/C$ in the quantizer. As a result, the gain along the signal path is still unity. Hence the input signal is still absence from the outputs of both integrators. Figure 11 shows output swings of both integrators by post layout simulation. Input signal frequency is 3.5 kHz with amplitude of 1.6 Vpp (2 Vpp reference voltage). No signal component is observed. Swings of both integrators are kept within ± 0.18 and ± 0.20 V respectively. Another interesting thing is that C_{P3} has no explicit influence on the modulator characteristic, while this is not the truth in the traditional design [6, 18]. In [6] analysis shows that the parasitic of the comparator will increase the order of the modulator and careful design and verification should be made to prevent the instability.

The parasitic does introduce an equivalent gain of $(C + C_{p1})/C$ to the second integrator. All the switches in Fig. 10 are complementary. The pMOS and nMOS transistors are with the same size of 7.2/0.18 μm. By simulation and post layout extraction, C_{p1} is found to be 28.8 fF, in which transistors contribute 23.6 fF and

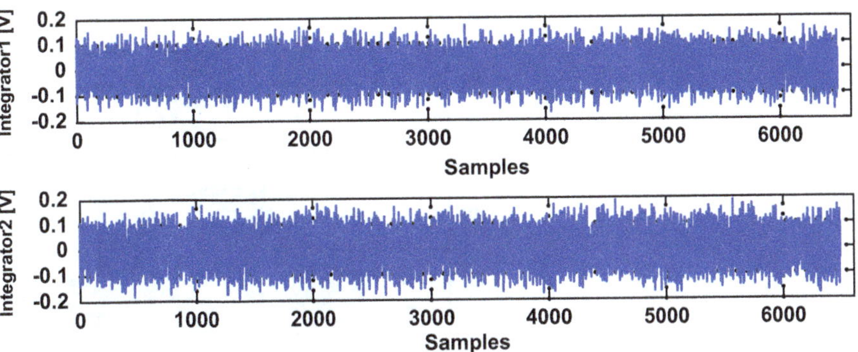

Fig. 11 Post layout simulation of integrators' output swings

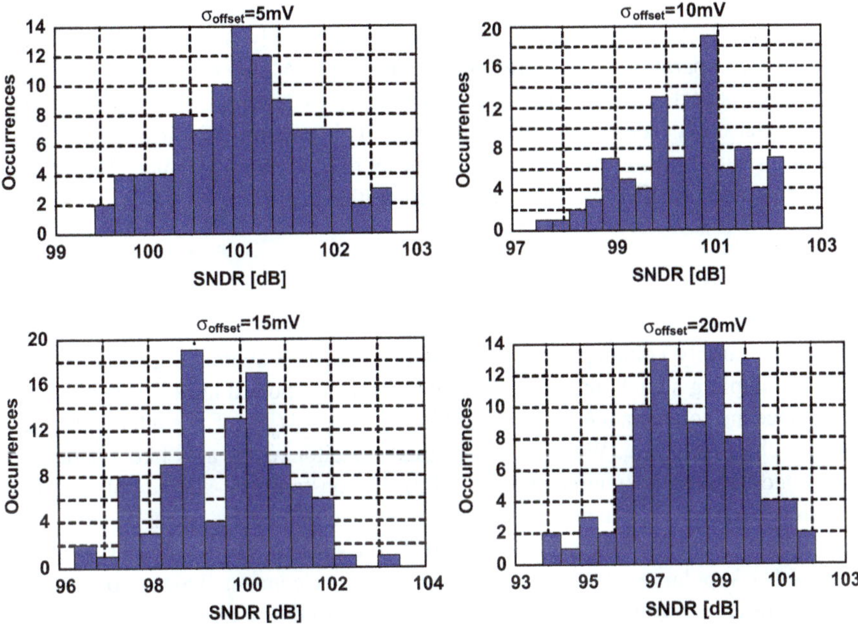

Fig. 12 Behavior simulation considering the comparator offset

interconnect provides 5.2 fF. The gain factor is approximately 1.28. Previous simulation shows that the performance of our modulator varies little even if $a_3 = 1.5$ and hence such a gain factor can be tolerated.

The essence of placing a multi-bit quantizer in the loop is to reduce the quantization step and thus quantization noise power can be decreased. An ideal flash quantizer has uniform decision interval. In the circuit implementation, however, the comparator offset will bring turbulence to each decision level. Some intervals are enlarged which leads to the quantization noise increment. Due to the mixture effect by the non-linear gain the modulator performance will be degraded. Figure 12 shows the behavior simulation of the modulator performance considering the

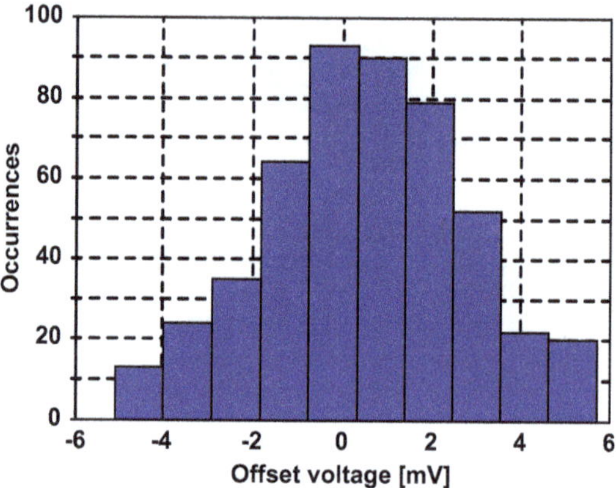

Fig. 13 Monte Carlo simulation of the comparator offset

comparator offset and non-linear gain effect simultaneously. According to the simulation, the comparator offset should be within 10 mV. Monte Carlo analysis of the comparator is performed in circuit level simulation to determine the offset voltage. A parameter σ is needed during the simulation. This parameter indicates the spread of process variations. In the simulation, σ is set to 3. Figure 13 depicts the simulation result, and the offset voltage can be kept within 6 mV.

Switches

Under low supply voltage, the conductance of the MOS switch is greatly limited. For the sampling switch, the bootstrapped technique reported in [22] is hence used to enhance the conductance as shown in Fig. 14. When M_6 is off, the internal capacitor C is charged to V_{DD}. When M_6 is on, this capacitor which acts like a battery is connected between the source and gate of M_6. A constant gate-source voltage (V_{gs}) is maintained and a constant conductance of M_6 is obtained. It should be noted that the gate parasitic of M_6 degrades V_{gs} due to charge sharing. To prevent this problem, C should be kept large. pMOS and nMOS switches are connected to the positive and negative reference voltages which are 1 and 0 V respectively. Switches connected to the virtue ground are all nMOS transistors because of relative low and stable VCMI of the amplifier. For output chopper circuits, complimentary switches are employed to transfer varying signal.

2.3 Decimation Filter

The block diagram of the decimation filter is shown in Fig. 15. The decimator is composed of five comb filters cascaded by two half-band filters [23]. The multistage design relaxes every sub-filter design and the overall hardware cost is quite smaller compared to the single stage realization. Each stage provides a decimation factor of two.

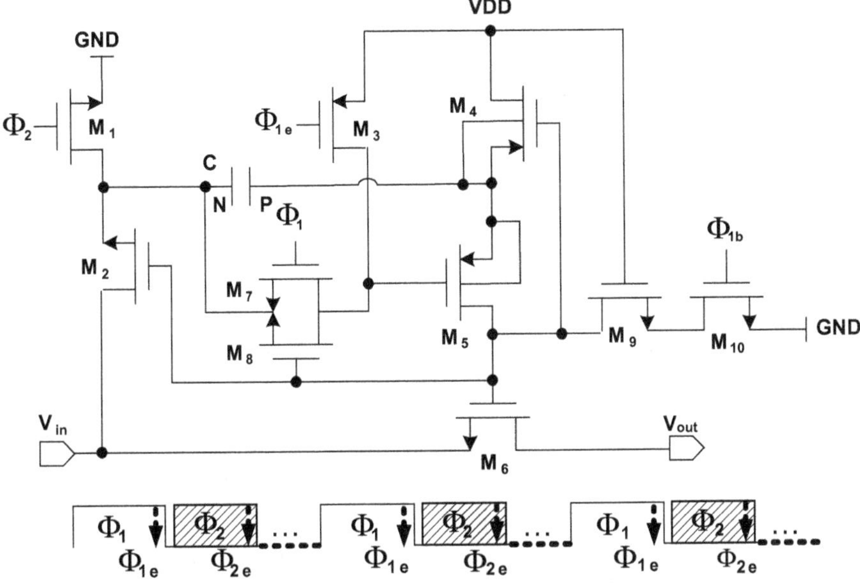

Fig. 14 Bootstrapped switch

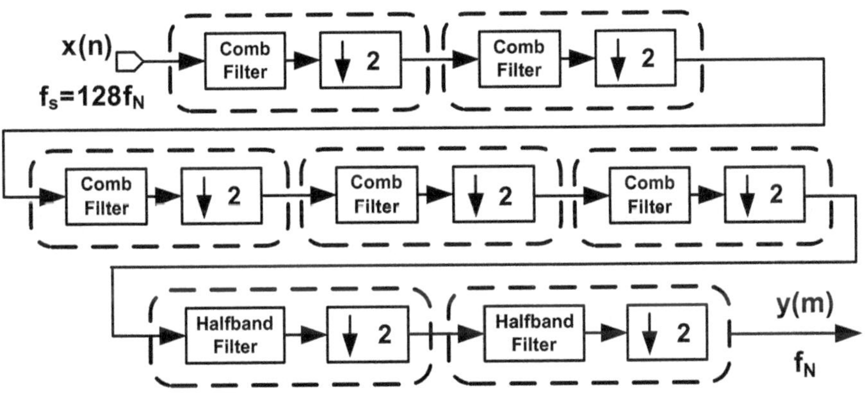

Fig. 15 Block diagram of the decimation filter

When the signal is processed by the first four stages, since the sampling rate is higher, the low pass filter only needs to provide enough attenuation around $z = -1$. Hence simple comb filters with a transfer function of $(1 + z^{-1})^4$ can be employed. For the last two stages, filters must have steep edge between the pass band and the stop band because the frequency range in which the noise will be folded into the baseband is close to the baseband. However previous stages have filtered out large portion of the noise, the last two stages do not need strong attenuation. The half-

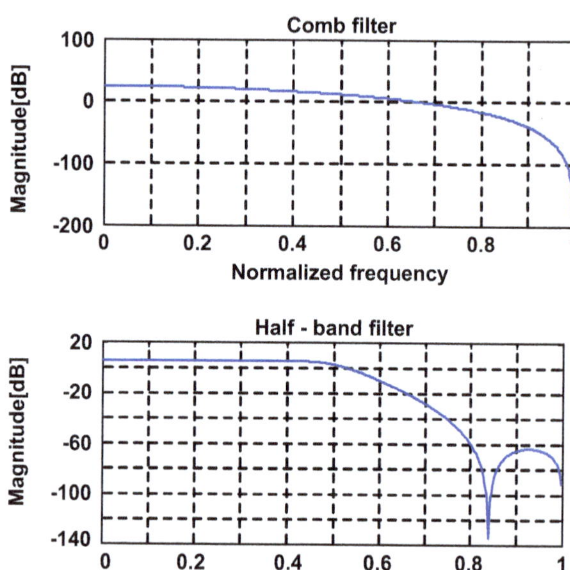

Fig. 16 Frequency response of sub filters

band filter adopts IIR type with the transfer function shown in (23). The magnitude response of sub filters is plotted in Fig. 16.

$$
H_{HB} = \frac{\dfrac{1}{8}z^{-5} + \dfrac{9}{16}z^{-4} + \dfrac{137}{128}z^{-3} + \dfrac{137}{128}z^{-2} + \dfrac{9}{16}z^{-1} + \dfrac{1}{8}}{\dfrac{9}{128}z^{-4} + \dfrac{88}{128}z^{-2} + 1}
\tag{23}
$$

Sub-filters adopt poly-phase topology so that they can work under low clock frequency. With sampling rate lowered down, the data refreshing rate in the sub filter gets slower compared to f_s. Especially for the half band filters in the last two stages, the data rates are only $f_s/64$ and $f_s/128$ respectively.

In the filter implementation, f_s is chosen as the master clock and for each stage, computation such as accumulation or shift can be distributed in time by multiplexing very few arithmetic units and hardware cost can be reduced [24]. To control the data flow paths, a multiplexer network is used. To illustrate the above idea, the hardware implementation of the fourth stage comb filter is presented here. Time domain relation between the comb filter's input x(n) and output y(n) is

$$
\begin{aligned}
y(n) &= x(n) + 4x(n-1) + 6x(n-2) \\
&\quad + 4x(n-3) + x(n-4)
\end{aligned}
\tag{24}
$$

During each data refreshing period, the filter needs to complete all the accumulation. Since output data rate of the fourth comb filter is only $f_s/16$, 16 master clock periods are hence available. Figure 17 shows the hardware scheme of the comb filter. Five-stage registers are used to generate time delay of the input data. For

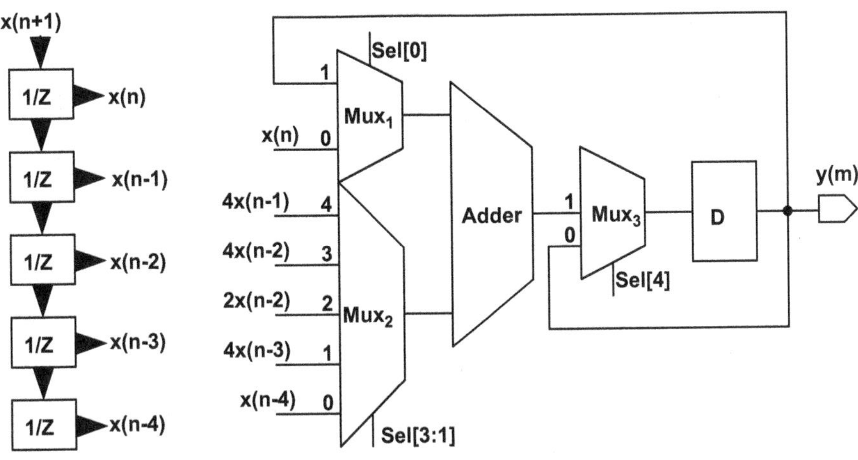

Fig. 17 Data paths of the fourth stage comb filter

arithmetic operation, only a single adder is required to finish all the summation in (24). The data paths can be configured by programming the control signal of the multiplexer. The detail operations are as follows:

- Step 1: x(n) and x(n − 4) are summed and the result is stored.
- Step 2: $4 \times (n-3)$ is accumulated to the previously stored result. Multiplying by 4 can be obtained by shifting x(n − 3).
- Step 3: $6 \times (n-2)$ should be accumulated. However directly multiplying by 6 is not easy to realize. So in this step only $2 \times (n-2)$ is added up.
- Step 4: $4 \times (n-2)$ is accumulated.
- Step 5: $4 \times (n-1)$ is summed to complete all the operations.

The half-band filter can be implemented in the same way. The only difference is the relatively more complex data paths control.

2.4 *Experiments Results*

The prototype ADC is fabricated in 0.18 μm 1P6M CMOS technology. Only regular threshold MOS transistors are adopted in the design. The chip microphotograph is shown in Fig. 18. The active die area excluding the bonding pads and decoupling capacitors are 0.5 mm^2. The modulator occupies 0.3 mm^2 and the decimator occupies 0.2 mm^2.

The test chip is directly bonded to the printed circuits board (PCB). The input signal is generated by an ultra low distortion signal generator (SG) DS360. Such equipment can provide very pure sine wave. When the amplitude of the output signal is smaller than 1.26 V, the noise floor of the equipment is less then $15 nV / \sqrt{Hz}$ which leads to integrated noise of only 1.9 μVrms with 16 kHz bandwidth. The

Fig. 18 Chip microphotograph

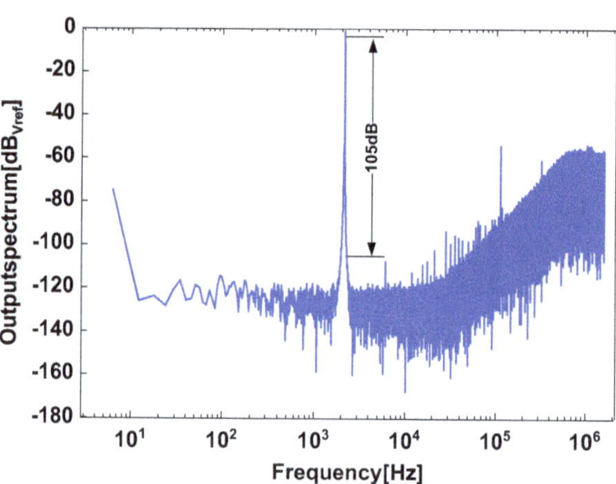

Fig. 19 Measured output spectrum of the modulator

total harmonic distortion (THD) of the output signal within 5.0 kHz is less than -110 dBc. The common mode voltage of the SG is 0 V when the SG is configured into balance mode. However the ADC under test requires a common mode voltage of 0.5 V, and 94 μF dc blocking capacitors are connected between SG and the chip.

The ADC works under 1 V supply. The reference voltages are chosen to be 1 and 0 V (2 Vpp) in order to maximize the input signal range. Figure 19 shows the output spectrum of the modulator under 32 kHz Nyquist sampling rate. The frequency of the input sine wave is 1.997 kHz and the amplitude is 1.68 Vpp which is 1.5 dB below the reference voltage. The tone is observed at the chopping frequency which shows the effectiveness of the chopper. Figure 20 depicts the output spectrum after

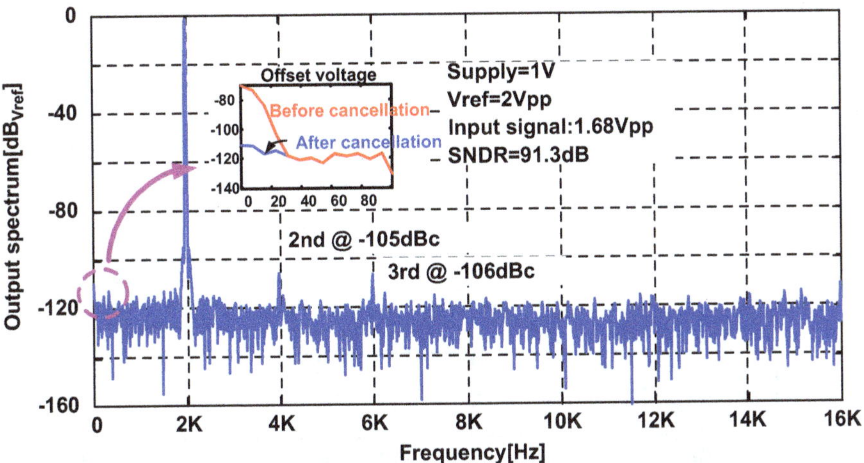

Fig. 20 Measured output spectrum of the ΔΣ-ADC

decimation. The measured peak SNDR across 20 Hz to 16 kHz is 91.3 dB. Thanks to the input feed forward path, the ADC exhibits an excellent linearity. The second and third harmonics are below -105 and -106 dBc respectively. Because the amplifier is chopped, flicker noise and offset introduced by the first stage amplifier is eliminated, however there is still a residue offset voltage because of reasons such as unsymmetrical charge injection to the first integrator, unsymmetrical clock feed through, non-zero odd order non-linear coefficients of the amplifier gain, offset of the second integrator, the quantizer offset and so on. For audio applications, the offset is not important which is different from instrumentation applications.

This offset voltage can be measured by shortening the inputs of the ADC. After measurement, the offset can be subtracted from the digital output. The spectrum near dc before and after offset cancellation is also shown in Fig. 20.

The ADC can adapt 8, 16 and 32 kHz Nyquist sampling rates by programming the clock frequency. Figure 21 shows the SNDR performance versus input amplitude at different sampling rates. The measured DR of the ADC is 93 dB. Power consumption is obtained by measuring the current drawn from the power supply. According to the measurement, the analog modulator consumes 190 μW and the decimator consumes 170 μW.

Performances of the ADC are summarized and compared to previous results in literatures as shown in Table 4. Two definitions of figure-of-merit as expressed in (25) and (26) are employed.

$$FOM_1 = \frac{P}{2^{ENOB} \times 2BW} \tag{25}$$

$$FOM_2 = \frac{P}{2^{2ENOB} \times 2BW} \tag{26}$$

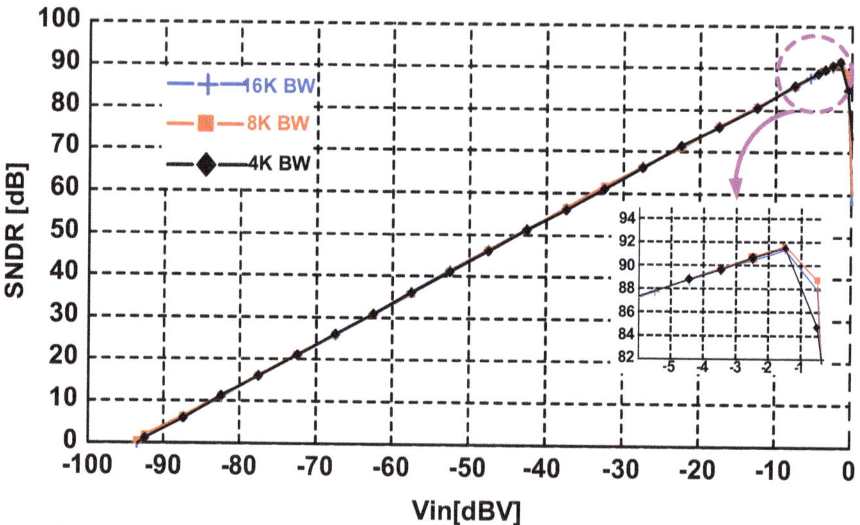

Fig. 21 Measured SNDR versus input amplitude

Symbol P denotes the power consumption, ENOB means the effective number of bits which is given by (SNDRpeak (in decibels)-1.76)/6.02 and BW is the signal bandwidth. Although (25) is the most popular, it is pointed out in [29] that definition in (26) is more suitable in comparisons among high precision $\Delta\Sigma$ modulators and ADCs. Our modulator achieves a very low FOM_1 and the best FOM_2 because of several reasons. The CIFB topology with feed forward path and multi-bit quantizer is adopted. Besides, efforts are made on the NTF design. All these techniques make a high precision modulator with low gain and low power amplifier possible. The analog summing circuitry proposed in this work overcomes the signal attenuation problem with the traditional circuitry. Hence additional amplification is prevented and power is saved. The decimator doubles the total power consumption and area cost. If advantage technology can be adopted, area of the decimator can shrink greatly and power dissipation of the decimator can be lowered down.

3   Ultra-Low Power DAC Design Principle

The digital-to-analog converter (DAC) not only converts the digital inputs into their analog waveforms but also needs to deliver efficient power to the load. The smaller the load impedance is the more power it requires. Unlike the design of ADC in which the power consumption of the source has already been excluded, in the design of DAC, the loading effect should be taken into consideration. As a result when the power consumption of DAC is measured, the power dissipation of the load is also included. Therefore, it is not logical to use the absolute power consumption to evaluate if a DAC is power efficient or not. A more reasonable standard is to use

Table 4 Performance summary and comparisons

References	[5]	[6]	[7]	[8]	[25]	[26]	[27]	[28]	This work
Process (μm)	0.18	0.13	0.09	0.18	0.18	0.18	0.13	0.18	0.18
Supply (V)	0.7	0.9	1	0.7	1	0.5	0.9	1.0	1
SNDR (dB)	95	73	80	81	62	74	89	84	91.3
DR (dB)	100	83	88	85	66	76	92	88	93
BW (kHz)	25	20	20	20	20	25	24	20	16
Power (μW)	870	60	140	36	42	740	1500	660	190/360 (with decimator)
Core area (mm^2)	2.16	0.42	0.18	0.72	0.54	0.6	1.44	1.57	0.3/0.5 (with decimator)
FOM_1 (pJ/step)	0.37	0.41	0.43	0.098	1.02	3.61	1.35	1.27	0.21/0.40 (with decimator)
FOM_2 (fJ/step)	0.0082	0.11	0.052	0.011	0.99	0.88	0.059	0.098	0.0066/0.012 (with decimator)

the parameter of power efficiency. A DAC with higher efficiency means that when delivering the same amount power into the load, the power consumption of itself is small.

The DAC in voice or audio applications is such kind of example, in which the DAC needs to drive very small load impedance and delivers sufficient power so that the sound level at the receiver end is loud enough. Beside large power delivering capability, such kind of DAC should have very large dynamic range which implies that the noise level should be very low. The oversampling DAC which allows tradeoff between speed and accuracy is hence popular in this specific application. The oversampling DAC is composed of a digital interpolator, a digital delta-sigma modulator and an analog power amplifier.

The interpolator increases the sampling rate. When digital audio signal passes through this block, its sampling rate is changed significantly. After interpolation the sampling rate becomes fast while the digital word length is still the same as its original input. A subsequent $\Delta\Sigma$ modulator can truncate the digital word with length of 16-bit or more into only one or a few bits. The noise shaping ability of the modulator allow the fast data stream maintain high SNR in the useful signal bandwidth. The analog reconstruction filter removes the out of band noise and finally clear analog waveform is restored. The power amplifier delivers the signal as well as noise power into the load. The receiver load has low pass frequency response and can remove the excessive noise out of the signal band. The power amplifier consumes most of the power in the DAC and the power efficiency of this amplifier is important to the overall efficiency. As a result, class-D amplifier which can achieve ideally 100 % power efficiency is preferred. However fast switching in Class-D amplifier may cause distortion and raise the noise floor, to improve the performance dedicated design technique should be adopted to lower down the switching rate. The power of digital interpolator and modulator should also be optimized. For the interpolator, cascaded architecture instead of single stage topology is adopted. Such topology costs much fewer hardware than that of the single stage. Besides, since the data rate varies significantly among different stages, if the highest clock rate is used as master clock, the arithmetic resource element can be multiplexed. This method can greatly save the hardware cost and hence save the power consumption.

In the next section, low power DAC employing the above techniques is described.

4 Ultra-Low Power DAC Design Example

This section presents the design and implementation of a single chip audio delta-sigma $\Delta\Sigma$-DAC with headphone driver for hearing aid. It consists of a digital interpolation filter, a $\Delta\Sigma$ modulator, a digital low-pass filter (DLPF) encoder and a class D output stage. A structure of chain of integrators with feedforward summation and local resonator feedbacks is adopted for $\Delta\Sigma$ modulator to shape quantization noise. Differing from the basic structure of $\Delta\Sigma$ DAC, a class D output stage playing

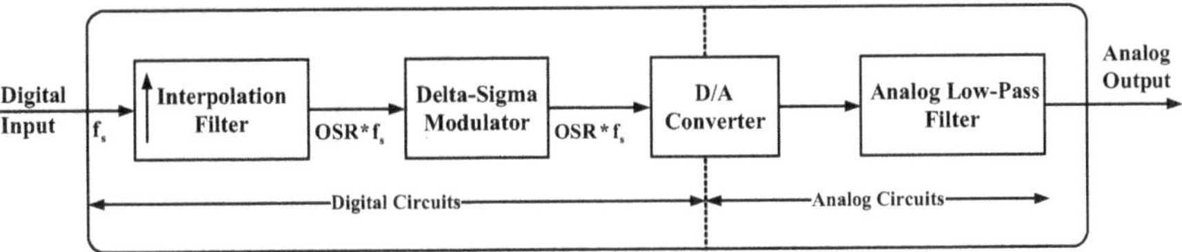

Fig. 22 Basic block diagram of a ΔΣ DAC

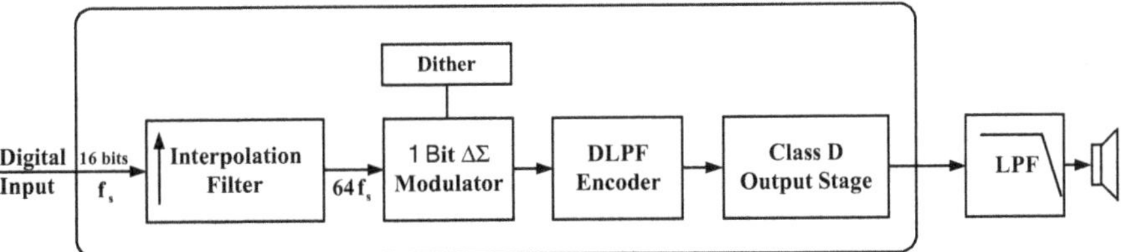

Fig. 23 Proposed architecture of the ΔΣ DAC

the role of headphone driver instead of the internal D/A converter is adopted. It is implemented in a 0.18 μm standard CMOS process, the die area is 2.46 mm². The ΔΣ DAC with headphone driver achieves 85.6 dB peak SNDR and 90 dB dynamic range (DR) and it can delivers power of 4.3 mW into a 376 Ω hearing aid headphone load.

The ΔΣ DACs have been widely used in high resolution applications, such as high quality digital audio signal processing and multi-media signal processing, because the use of oversampling techniques overcomes the analog accuracy problem by trading digital complexity and speed for the desire insensitivity to analog nonidealities, and allows the use of robust and simple analog circuitry [30]. So, research on high resolution data converters pays more and more attention to this structure recently. The basic block diagram of ΔΣ DAC is shown in Fig. 22 it consists of a digital interpolation filter, a ΔΣ modulator, an internal D/A converter and an analog low-pass filter. Some designs use improved basic structure [31].

The structure of the internal D/A converter is very important, 1-bit DAC will be very simple, but the high slew rate of the 1-bit DAC output signal and the large amount of out-of-band noise power it contains, make the design of the subsequent analog smoothing filter a difficult task [30]. At the same time, a multi-bit DAC will require additional circuitry for the filtering or cancellation of the DAC nonlinearity error, which results in a more complex DAC. Furthermore, if the DAC want to get higher load driving ability, an additional power driver is needed. Comparing with class AB output stage used in [32, 33], class D output stage works in switch mode, so it is able to directly couple to digital audio signal without the need to convert the digital signal to analog first. It means that we can complete the entire signal processing before the analog low-pass filter in digital domain. Following the ΔΣ modulator, as a hardware-efficient method, a class D output stage playing the role of headphone driver will result in a simple circuit structure. The proposed architecture of the ΔΣ DAC with headphone driver is shown in Fig. 23. The interpolation filter

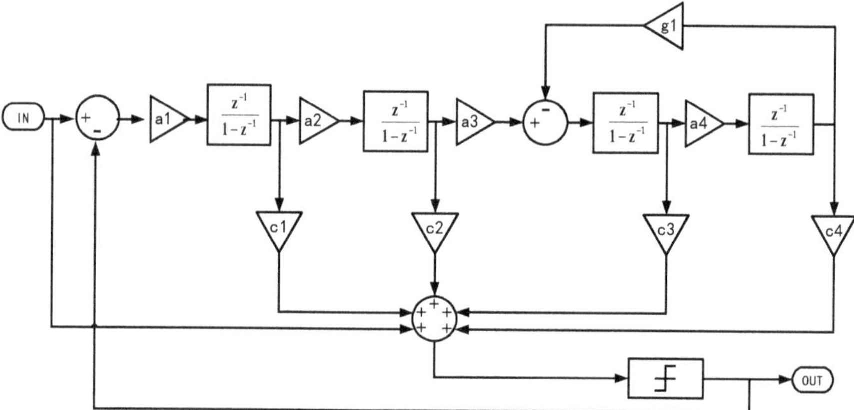

Fig. 27 Structure of the $\Delta\Sigma$ modulator

Table 5 Modulator coefficients

Coefficients	Value	Coefficients	Value
a_1	1/8	c_1	2
a_2	1/4 + 1/8 + 1/32	c_2	2
a_3	1/4 + 1/16	c_3	1
a_4	1/16 + 1/32	c_4	2
g_1	1/64		

$$NTF(z) = \frac{A_4 z^4 + A_3 z^3 + A_2 z^2 + A_1 z + A_0}{B_4 z^4 + B_3 z^3 + B_2 z^2 + B_1 z + B_0} \tag{28}$$

where

$A_4 = 1$, $A_3 = -4$, $A_2 = 6 + a_4 g_1$, $A_1 = -4 - 2 a_4 g_1$, $A_0 = 1 + a_4 g_1$, $B_4 = 1$, $B_3 = -4 + a_1 c_1$, $B_2 = 6 + a_4 g_1 - 3 a_1 c_1 + a_1 a_2 c_2$, $B_1 = -4 - 2 a_4 g_1 + 3 a_1 c_1 - 2 a_1 a_2 c_2 + a_1 a_4 g_1 + a_1 a_2 a_3 c_3$, $B_0 = (1 + a_4 g_1)(1 - a_1 c_1 + a_1 a_2 c_2) + a_1 a_2 a_3 (a_4 c_4 - c_3)$.

The main difference between this structure and the conventional CIFF structure is the local resonator feedback around pairs of integrators in the loop filter, which moves the open-loop poles away from DC along the unit circle. And a notch in the spectrum caused by the resonator is shown in Fig. 28 (considering a modulated 20 bits sine wave whose frequency is 3750 Hz, amp is 0.8 V), which will realize a better result of noise shaping and reach a higher SNDR.

The coefficients of the designed modulator are summarized in Table 5. They are set according to the simulation results and the principle of easy hardware implementation.

What's more, in our design a dither is added to the resonator before the quantizer of the $\Delta\Sigma$ modulator to suppress the idle tones. It is generated using a linear feedback shift register (LFSR) in Galois configuration. The generator polynomial of the LFSR is $x^{19} + x^{18} + x^{17} + x^{14} + 1$.

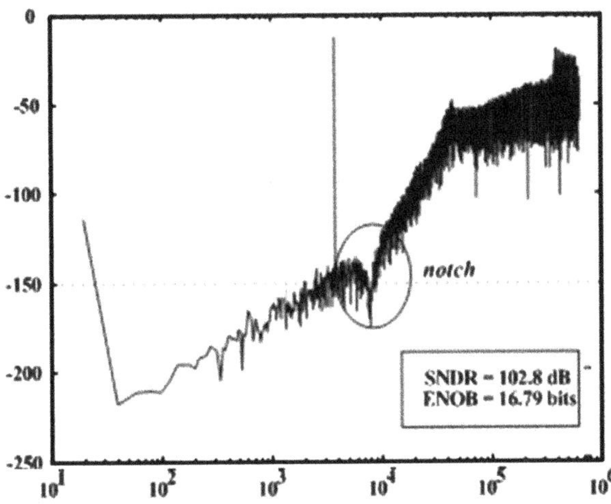

Fig. 28 PSD of the output of the $\Delta\Sigma$ modulator

Table 6 Signal and value of DLPF

Signal	Value			
DIN (n)	0	0	1	1
DIN (n − 1)	0	1	0	1
S	0	1	1	0
C	0	0	0	1
EPOUT12-EPOUT34	1		0	−1

4.3 DLPF Encoder

Typically, the 1 bit output of the $\Delta\Sigma$ modulator will control the switching activity of the class D output stage directory [37]. A DLPF encoder is placed after the $\Delta\Sigma$ modulator, whose transfer function is $H(z)=0.5\,(1+z^{-1})$, shown in Fig. 29. It will result in a three state output $(-1, 0, 1)$.

S is the sum of the adder of the encoder, and C is the carry of the adder of the encoder. There are three combinations of S and C from Table 6, which is 00, 10 and 01. Let DIN (n) be 65536 points of modulated result of a random sine wave, and get the statistical result of SC, shown in Fig. 30. The probability of SC = 10 is the biggest, so the output of the class D output stage is set to zero when SC = 10.

4.4 Class D Output Stage

In order to get a higher output, the full-bridge topology is adopted as the power output stage, which consists of two sets of switches. A full-bridge topology is suitable for open loop design, and the different output structure of the bridge topology inher-

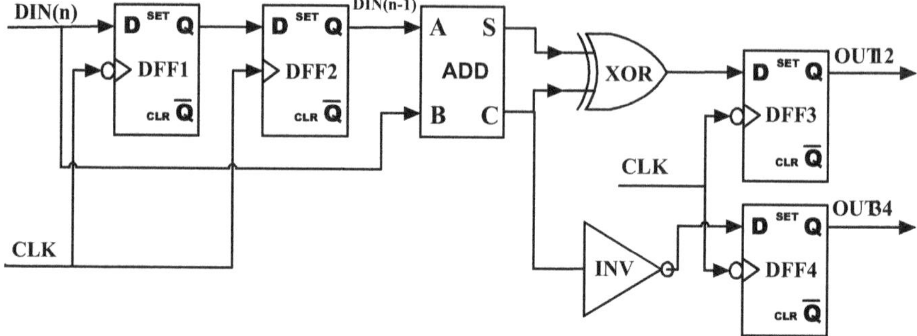

Fig. 29 Proposed DLPF encoder

Fig. 30 Statistical results of SC

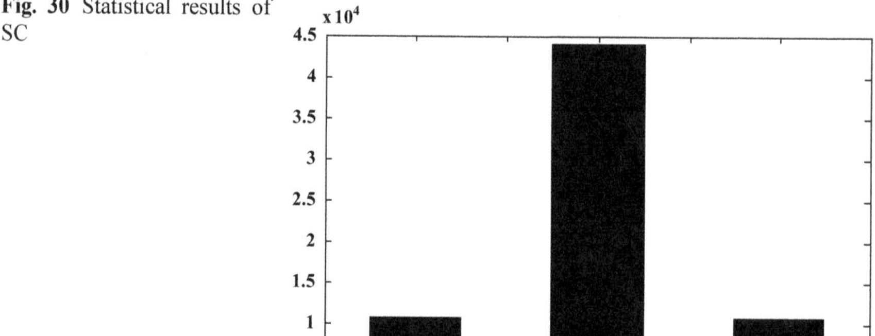

ently can cancel the even order of harmonic distortion components and DC offsets. Figure 31 shows the hardware configuration of the proposed class D output stage.

Dead-time control is very important for a class D amplifier, because once both high and low side MOSFETs are turned on simultaneously, a low resistance between power supply and ground will result in a large current, which can cause significant energy loss and a crossover-type distortion or even make the MOS be shoot though. In this design we should ensure that the signals in each branch of the differential path have equal rise and fall times, weak cross-coupled inverters are inserted between the two differential lines of the output stage. The feedforward provided by the cross-coupled inverters helps minimizing the signal skew [38].

4.5   Measurement Results

The prototype $\Delta\Sigma$ DAC with headphone driver has been fabricated in 0.18 μm 1P6M CMOS process with only regular threshold voltage NMOS and PMOS. A microphotograph highlighting its main parts of the chip is shown in Fig. 32. In order

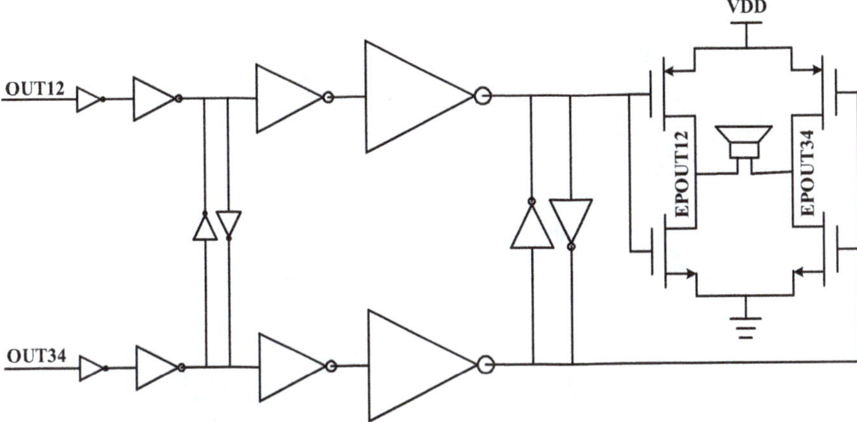

Fig. 31 The Class-D output stage

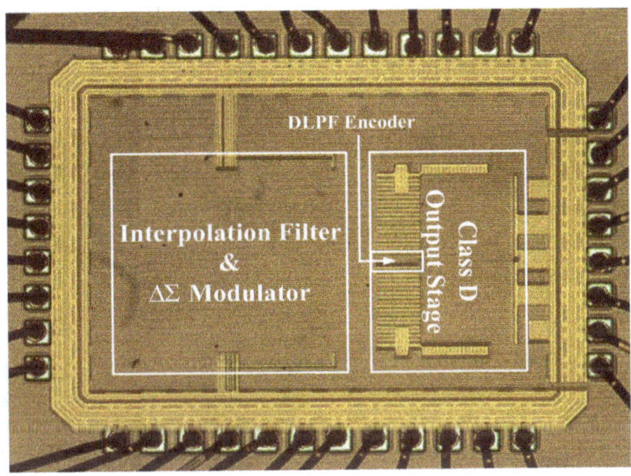

Fig. 32 Microphotograph of Chip

to get a high efficiency of the class D output stage we need to design the H bridge of the class D output stage in an extremely large W/L ratio. In [39] two methodologies are proposed, in order to get a matching of output transistor and an adequate distribution of current the waffle layout method is adopted

Considering a 3.414 kHz input sine wave with amplitude of 9.6 dB below full-scale (FS). Figure 33. shows the spectrum of measured 65536 points Rife-Vincent1-windowed FFTs of the delta-sigma DAC output. The measured DR and peak SNDR in the 20 Hz~10 kHz bandwidth is 90 dB and 85.6 dB respectively. The die area is 2.46 mm^2. The key performance is summarized in Table 7.

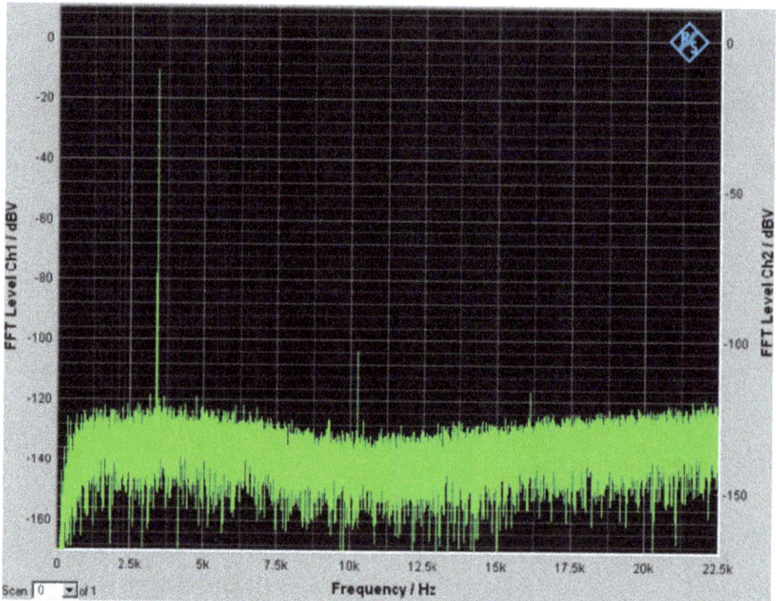

Fig. 33 Measured output spectrum

Table 7 Performance summary

Process	0.18 μm standard CMOS
Supply voltage	1.0-V
Signal bandwidth	20 Hz~10 kHz
Clock frequency	1.28 MHz
OSR	64
Output swing	0.46 Vpp
Load	Hearing aid headphone (376 Ω)
Peak SNDR	85.6 dB
Dynamic range (A-weight)	90 dB
Power dissipation	4.3 mW
Chip area	2.46 mm^2

References

1. Y. Geerts, M. Steyaert, and W. Sansen, *Design of Multi-Bit Delta-Sigma A/D Converters.* Norwell, MA: Kluwer, 2002.
2. Liyuan Liu, Dongmei Li, Liangdong Chen, Yafei Ye and Zhihua Wang, "A 1-V 15-bit Audio $\Delta\Sigma$-ADC in 0.18 μm CMOS," *IEEE Trans. Circuits Syst. I, Reg. Papers*, vol. 59, pp. 915–925, May. 2012.
3. P. Benabes, A. Gauthier, and D. Billet, "New wideband sigma-delta convertor," *IET Electron. Lett.*, vol. 29, pp. 1575–1577, Aug. 1993.
4. J. Silva, U.-K. Moon, J. Steensgaard, and G.C. Temes, "Wideband low-distortion delta-sigma ADC topology," *IET Electron. Lett.*, vol. 37, pp. 737–738, Jun. 2001.

5. H. Park, K. Y. Nam, D. K. Su, K. Vleugels, and B.A. Wooley, "A 0.7-V 870-μW digital-audio CMOS sigma-delta modulator," *IEEE J. Solid-State Circuits*, vol. 44, pp. 1078–1088, Apr. 2009.

6. J. Roh et al., "A 0.9-V 60-μW 1-bit fourth-order delta-sigma modulator with 83-dB dynamic range," *IEEE J. Solid-State Circuits*, vol. 43, pp. 361–370, Feb. 2008.

7. L. Yao, M. Steyaert, and W. Sansen, "A 1-V 140-μW 88-dB audio sigma-delta modulator in 90-nm CMOS," *IEEE J. Solid-State Circuits*, vol. 39, pp. 1809–1818, Nov. 2004.

8. Y. Chae, I. Lee, and G. Han, "A 0.7 V 36 μW 85 dB-DR audio ΔΣ modulator using class-C inverter," in *IEEE ISSCC Dig. Tech. Papers*, Feb. 2008, pp. 490–491.

9. S. R. Norsworthy, R. Schreier, and G. C. Temes, *Delta-Sigma Data Converters:Theory, Design and Simulation*. New York: IEEE Press, 1997.

10. H. Zare-Hoseini and I. Kale, "On the effects of finite and non-linear DC-gain of the amplifiers in switched-capacitor ΔΣ modulators," in *Proc. 2004 IEEE Int. Symp. Circuits and Systems (ISCAS'04)*, May 2004, pp. 2547–2550.

11. M. Yavari and A. Rodriguez-Vazquez, "Accurate and simple modeling of amplifier dc gain nonlinearity in switched-capacitor circuits," in *Proc. IEEE 18th European Conf. on Circuit Theory and Design (ECCTD'07)*, Aug. 2007, pp. 144–147.

12. K. Abdelfattah and B. Razavi, "Modeling op amp nonlinearity in switched-capacitor sigma-delta modulators," in Proc. 2006 IEEE Custom Integrated Circuits Conf.(CICC'06), Sep. 2006, pp. 197–200.

13. F. -C. Chen and C.-L Hsieh, "Modeling harmonic distortions caused by nonlinear op-amp dc gain for switched-capacitor sigma-delta modulators," IEEE Trans. Circuits Syts. II, Exp.Briefs, vol. 56, pp. 694–698, Sep. 2009.

14. Y. Yang, A. Chokhawala, M. Alexander, J. Melanson and D. Hester, "A 114-dB 68-mW chopper-stabilized stereo multibit audio ADC in 5.62 mm^2," *IEEE J. Solid-State Circuits*, vol. 38, pp. 2061–2068, Dec. 2003.

15. R. T. Baird and T. S. Fiez, "Linearity enhancement of multibit delta-sigma A/D and D/A converters using data weighted averaging," IEEE Trans. Circuits Systems II, Analog Digit. Signal Process., vol. 42, pp. 753–761, Dec. 1995.

16. O. Choksi and L.R. Carley, "Analysis of switched-capacitor common-mode feedback circuit," *IEEE Trans. Circuits Systems II, Analog Digit. Signal Process.*, vol. 50, pp. 906–917, Dec. 2003.

17. C.C. Enz and G. C. Temes, "Circuit techniques for reducing the effects of op-amp imperfections: Autozeroing, correlated double sampling, and chopper stabilization," *Proc. IEEE*, vol. 84, pp. 1584–1614, Nov 1996.

18. L. Yao, M. Steyaert, and W. Sansen, "A 1-V, 1-MS/s, 88-dB sigma-delta modulator in 0.13-μm digital CMOS technology,"in *IEEE Symp. VLSI Circuits Dig. Tech. Papers*, Jun. 2005, pp. 180–183.

19. R. Gaggl, M. Inversi, and A. Wiesbauer, "A power optimized 14-bit SC ΔΣ modulator for ADSL CO applications," in *IEEE ISSCC Dig. Tech. Papers*, Feb. 2004, pp. 82–83.

20. S. Kwon and F. Maloberti, "A 14 mW multi-bit ΔΣ modulator with 82 dB SNR and 86 dB DR for ADSL2+," in *IEEE ISSCC Dig. Tech. Papers*, Feb. 2006, pp. 161–162.

21. L. Liu, D. Li, L. Chen, Y. Ye and Z. Wang, "A 1 V 15-bit audio ΔΣ ADC in 0.18 μm CMOS," in *Proc. 2011 IEEE Int. Symp. Circuits and Systems (ISCAS'11)*, May 2011, pp. 510–513.

22. M. Dessouky and A. Kaiser, "A 1 V 1 mW digital-audio ΔΣ modulator with 88 dB dynamic range using local switch bootstrapping," in *Proc. 2000 IEEE Custom Integrated Circuits Conf.(CICC'00)*, May 2000, pp. 13–16.

23. J. C. Candy and G. C. Temes, Oversampling Methods for A/D and D/A Conversion. New York: IEEE Press, 1992.

24. L. Liu, R. Chen and D. Li, "A 20-bit sigma-delta D/A for audio applications in 0.13 um CMOS," in *Proc. 2007 IEEE Int. Symp. Circuits and System (ISCAS'07)*, May 2007, pp. 3622–3625.

25. C. Su, P.-C. Lin and H. Lu, "An inverter based 2-MHz 42-μW ΔΣ ADC with 20-kHz bandwidth and 66 dB dynamic range," in *Proc. 2006 IEEE Asian Solid-State Circuits Conf.(ASSCC'06)*, Nov. 2006, pp. 63–66.

26. K.-P. Pun, S. Chatterjee, and P. R. Kinget, "A 0.5-V 74-dB SNDR 25-kHz continuous-time delta-sigma modulator with a return-to-open DAC," *IEEE J. Solid-State Circuits*, vol. 42, pp. 496–507, Mar. 2007.
27. M. G. Kim et al. "A 0.9 V 92 dB double-sampled switched-RC $\Delta\Sigma$ audio ADC," in IEEE Symp. *VLSI Circuits Dig. Tech. Papers*, Jun. 2006, pp. 160–161.
28. C.-H. Kuo, D.-Y. Shi, and K.-S. Chang, "A low-voltage fourth-order cascade delta-sigma modulator in 0.18-μm CMOS," *IEEE Trans. Circuits Syst. I, Reg. Papers*, vol. 57, pp. 2450–2461, Sep. 2010.
29. H.-S. Lee and C. G. Sodini, "Analog-to-digital converters: Digitizing the analog world," Proc. IEEE, vol. 96, pp. 323–334, Feb. 2008.
30. Schreier R., Temes G.C., *Delta-Sigma Data Converters*, Wiley-IEEE Press, 2005, pp. 219–222.
31. Ichiro Fujimori, Tetsuro Sugimoto, "A 1.5 V, 4.1 mW dual-channel audio delta-sigma D/A converter," IEEE J. Solid-State Circuits, vol. 33, no. 12, pp. 1863–1870, December 1998.
32. Kim-Fai Wong, Ka-Ian Lei, Seng-Pan U and R.P. Martins, "A 1-V 90 dB DR Audio Stereo DAC with embedding Headphone Driver," IEEE Asia Pacific Conference on Circuits and Systems, pp. 1160–1163, November-December 2008.
33. Steven R. Norsworthy, Richard Schreier, Delta-Sigma Data Converters: Theory, Design, and Simulation, Wiley-IEEE Press, 1996, pp. 413–415.
34. Kyehyung Lee et al. "A 0.8 V, 2.6 mW, 88 dB Dual-Channel Audio Delta-Sigma D/A Converter With Headphone Driver," IEEE J. Solid-State Circuits, vol. 44, no. 3, pp. 916–927, March 2009.
35. Run Chen, Liyuan Liu, Dongmei Li, "A cost-effective digital front-end realization for 20-bit $\Sigma\Delta$ DAC in 0.13 μm CMOS," IEEE Custom Integrated Circuits Conference, pp. 447–450, September 2007.
36. Jeongjin Roh, Sanho Byun, Youngkil Choi, Hyungdong Roh, Yi-Gyeong Kim, and Jong-Kee Kwon, "A 0.9-V 60-μW 1-Bit Fourth-Order Delta-Sigma Modulator With 83-dB Dynamic Range," IEEE J. Solid-State Circuits, vol. 43, no. 2, pp. 361–370, February 2008.
37. Kathleen Philips, John van den Homberg, Carel Dijkmans, "Power DAC: a single-chip audio DAC with a 70%-efficient power stage in 0.5 μm CMOS," IEEE International Solid-State Circuits Conference, pp. 154–155, February 1999.
38. Jorge Varona, Anas A. Hamoui, and Ken Martin, "A Low-Voltage Fully-Monolithic $\Delta\Sigma$-Based Class-D Audio Amplifier," ESSCIRC, pp. 545–548, September 2003.
39. Joseph S. Chang, Meng-Tong Tan, Zhihong Cheng, and Yit-Chow Tong, "Analysis and Design of Power Efficient Class D Amplifier Output Stages," IEEE Transactions on circuits and system, vol. 47, no. 6, pp. 897–902, June 2000.

Low Power Design Methodologies for Digital Signal Processors

Liming Chen, Yong Hei, Zenghui Yu, Jia Yuan and Jinyong Xue

1 Low Power Design Introduction for Digital CMOS Circuits

Before the introduction, it is necessary to differentiate power and energy first, especially for battery operated system. Power is the instantaneous power dissipation in the system, and energy is the integral of power over time. The power used by a given system varies over time depending on what it is doing, while it is energy that determines battery life.

At present, CMOS is the mainstream technology for digital VLSI. In this section we firstly introduce the main source of power consumption in CMOS circuits, and then give a briefly summary on low power design Methodologies.

1.1 Sources of Power Dissipation in CMOS Circuits

There are a number of sources of power consumption in digital CMOS circuits, which can be subdivided into **dynamic** and **static** power consumption. Dynamic power is the power consumed when the device is active, which means when signals are changing values. Static power is the power consumed when the device is powered up but no signals are changing value [1].

Dynamic Power

Dynamic power dissipation mainly consists of switching power and short circuit power, of which the former dominates. Switching power, known as capacitive

L. Chen (✉) · Y. Hei · Z. Yu · J. Yuan · J. Xue
ASIC and System Department, Institute of Microelectronics, Chinese Academy of Sciences,
Beijing, China
e-mail: chenliming@ime.ac.cn

N. N. Tan et al. (eds.), *Ultra-Low Power Integrated Circuit Design*,
Analog Circuits and Signal Processing 85, DOI 10.1007/978-1-4419-9973-3_5,

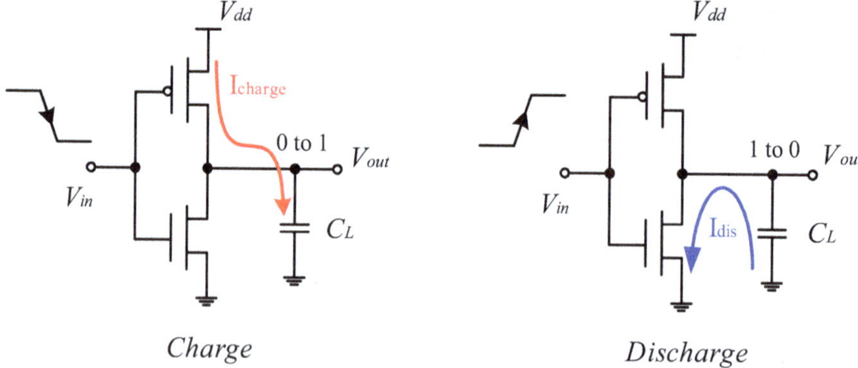

Fig. 1 Switching power of a CMOS inverter

power, is consumed during charging and discharging the parasitic load capacitance. Figure 1 illustrates switching power by the transition of a basic CMOS inverter.

The energy required per transition in a CMOS inverter is a fixed value, given by:

$$E = C_L \cdot V_{dd}^2 \tag{1}$$

Where C_L is the load capacitance and V_{dd} is the supply voltage.

If the frequency of the system clock is f, and there are N nodes in the system, then the switching power of the system can be modeled by the following equation:

$$P_{switch} = V_{dd}^2 \cdot f \cdot \sum_{i=1}^{N} \alpha_i \cdot C_i \tag{2}$$

Where C_i and α_i is the load capacitance and probability of output transition for node i.

If we define

$$C_{eff} = \sum_{i=1}^{N} \alpha_i \cdot C_i \tag{3}$$

We can also describe the switching power as below:

$$P_{switch} = C_{eff} \cdot V_{dd}^2 \cdot f \tag{4}$$

The power dissipation due to short-circuit current is caused by the inherent non-ideal characteristic of the input signals in CMOS circuits. Because of the finite slope of the input signal during switching, the PMOS and NMOS devices are simultaneously turned ON for a very short period of timeduring a logic transition, allowing a short-circuit current to run from V_{dd} to ground. Figure 2 shows the short-circuit current in a CMOS inverter.

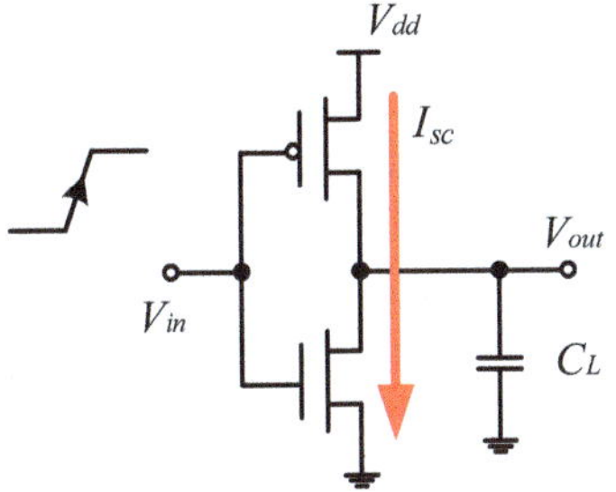

Fig. 2 Short circuit current of a CMOS inverter

Normally, short-circuit power is only a small part in dynamic power. At the overall circuit level, if rise/fall times of all signals are kept constant within a range, overall short-circuit current will be kept within bounds (maximum 10–15 % of the total dynamic power dissipation). In the extreme case, if $V_{dd} < V_{THn} + |V_{THp}|$, the short-circuit dissipation will be completely eliminated, because the devices are never ON simultaneously [2].

For the overall dynamic power is dominated by the switching power, we often simply use the switching power formula to compute the total dynamic power

$$P_{dyn} \approx P_{switch} = C_{eff} \cdot V_{dd}^2 \cdot f \tag{5}$$

Formula (5) is very important for power computation for CMOS digital integrated circuits. From the (5), we can draw conclusion that switching power is not a function of transistor size, resistance, charging/discharging current, but rather a function of supply voltage, clock frequency, switching activity and load capacitance. To reducing the dynamic power dissipation, there are many techniques can be used, which focus on the voltage, frequency, capacitance components of the equation, as well as reducing the data-dependent switching activity.

Static Power

In the digital CMOS circuits, static power, which is a constant factor and has nothing to do with the switching activity, is mainly caused by leakage current. There are four main sources of leakage currents in a CMOS device: Sub-threshold Leakage (I_{SUBTH}), Gate Leakage (I_{GATE}), Diode (drain-substrate) Reverse Bias Junction Leakage (I_{REV}), Gate Induced Drain Leakage (I_{GIDL}). Figure 3 illustrates the main leakage currents in a MOS device [1].

Amongst all the above listed leakage current components, sub-threshold leakage is the dominant component of leakage current in previous technology nodes. Sub-threshold leakage current flows from the drain to the source of a transistor when it is off (in fact not turned completely off). This happens in weak inversion mode when the applied voltage V_{GS} is less than the threshold voltage V_{TH} of the transistor. The value of sub-threshold leakage (I_{SUBTH}) current can be given by the following equation:

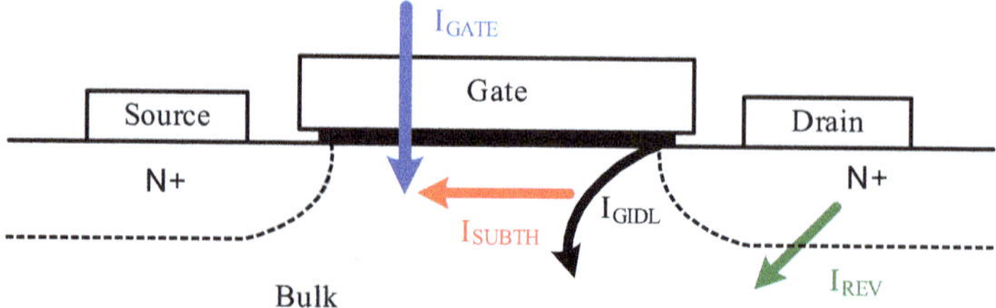

Fig. 3 Main leakage current in MOS device

$$I_{SUBTH} = \mu C_{ox} V_{T0}^2 \frac{W}{L} \cdot e^{\frac{V_{GS} - V_{TH}}{n V_{T0}}} \tag{6}$$

Where W and L are the dimensions of the MOS transistor, the n is a function of the device fabrication process and ranges from 1.0 to 2.5, and V_{T0} is the thermal voltage kT/q.

Sub-threshold leakage makes low power design rather complicated. For reducing dynamic power, people choose to decrease supply voltage V_{dd} continuously. Meanwhile, in order to maintain the good performance, we need to lower threshold voltage V_{TH} as we lower V_{dd}. But unfortunately, lowering V_{TH} will result in an exponential increase in the sub-threshold leakage current according to equation (6). Even worse, sub-threshold leakage current increases exponentially with temperature, which makes the leakage power easy to exceed the design constraint at the worst case while it is acceptable at room temperature.

There are several methods to deal with sub-threshold leakage current, one of them is using multi-threshold logic. For critical path in the circuit, low threshold cells are used to achieve the timing constraint, while for other paths, high threshold cells with lower sub-threshold leakage can be chosen. Another important technique is power-gating, which shut down the supply voltage parts of the chip when they are not working. Some other methods often used for lowering sub-threshold leakage are variable threshold CMOS, utilizing the stack effect, long channel devices, and so on.

Gate leakage, which occurs as a result of tunneling current through the gate oxide, is becoming significant in the sub-100 nm era. To maintain the current drive of the MOS transistor while scaling its horizontal dimensions, the gate oxide (SiO2) thickness is scaled as well. When the oxide thickness becomes of the order of just a few molecules in the sub-100 nm technology nodes, gate tunneling current can increase to the amount that may be comparable with sub-threshold leakage. Now, using high-k dielectric materials appears to be the only effective way of keeping gate leakage under control when scaling the gate thickness [2].

Reverse Bias Junction Leakage (I_{REV}), another leakage contribution that should not be ignored, is caused by minority carrier drift and generation of electron/hole pairs in the depletion regions. With the decreasing thickness of the depletion regions owing to the high doping levels, tunneling through narrow depletion region becomes an issue in sub-50 nm technology nodes.

Gate Induced Drain Leakage (I_{GIDL}) is the current flowing from the drain to the substrate induced by the high electric field effect in the MOSFET drain caused by a high V_{DG}. Generally speaking, I_{GIDL} plays a minor role and can be ignorable in most of today's designs.

Before 90 nm technology node, static power in digital CMOS circuits is usually very small compared to dynamic power. But in the sub-90 nm technology nodes, static power has become a big problem and should be paid much attention to, especially for the battery operated systems widely used today.

1.2   Basic Low Power Design Methodologies

Low-power design can be applied on different levels, such as the system level, architectural level, the logic/RT level, the gate level, the circuit/transistor level and the physical/technology level, et al. Using low power design techniques at different levels can approach different power savings. Generally, the high level design methods have more effect on power saving.

Power or energy optimization can also be performed at different stages in the design process and may address different targets such as dynamic power or static power. In the following, we will briefly introduce the main low power design methodologies realized at design time, runtime and standby separately. We should always be aware of the fact that low power design usually calls for employing a combination of various low power techniques.

Design Time Power-Reduction Techniques

Reducing the Supply Voltage

As mentioned in formula (5), dynamic power dissipation depends quadratically on the supply voltage V_{dd}. Voltage scaling is therefore the most attractive and effective method for low power design. However, when supply voltage is lowered, it comes at a cost: the delay of the CMOS logic gate increases. The loss of performance can be compensated by some architecture and logic optimizations, such as parallelism and pipelining [3].

Using parallel processing structure means build N ($N \geq 2$) same functional module instead of one, and then N problems can be solved concurrently. Therefore, each module can now operate at the 1/N speed with the system keeping the same throughput, and the clock frequency can be slowed down to 1/N. The relaxed delay requirement enables a reduction of the supply voltage without depressing the system performance [4]. Figure 4 gives an example of parallel design. The power

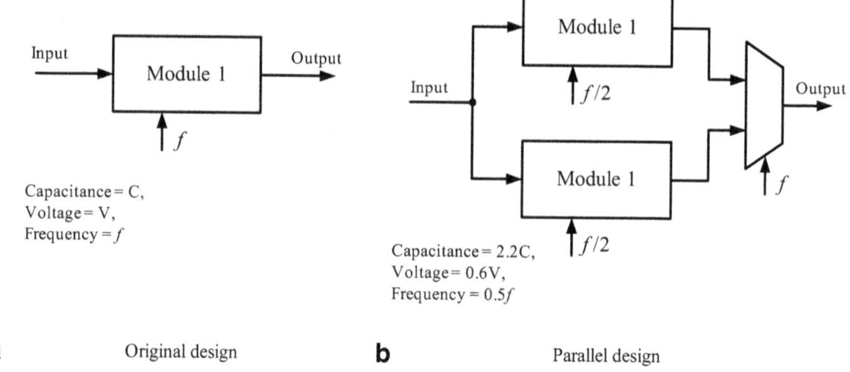

Fig. 4 Parallel structure

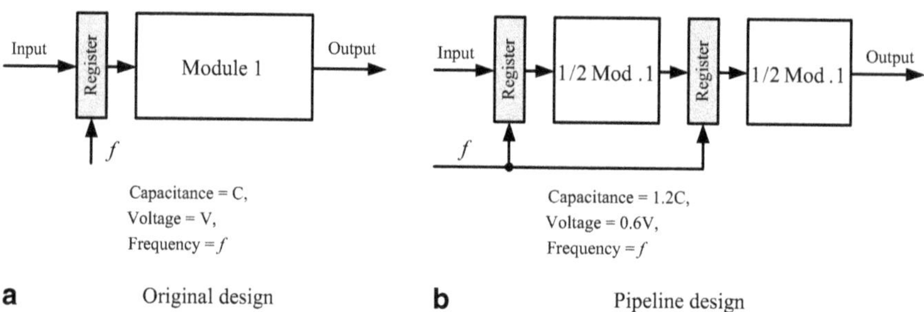

Fig. 5 Pipeline structure

consumption of the original design and the parallel design can be computed by the following equations:

$$P_{original} = C \cdot V^2 \cdot f \tag{7}$$

$$P_{parallel} = 2.2C \cdot (0.6V)^2 \cdot (0.5f) = 0.4P_{original} \tag{8}$$

Through the comparison of equation (7) and (8), we can see the effect of parallel structure in reducing power. The main shortcoming of this method is the substantial area overhead.

Pipelining is another method introducing concurrency. This technique divides the critical paths by inserting extra registers between the neighboring modules. Due to the reducing of delay, the system is now able to work at lower supply voltage with keeping the throughput. Figure 5 gives an example of using pipeline in a design. Compared to the fact that parallel structure usually increases the area by more than 100%, which is often not acceptable, the pipeline structure can gain about the same

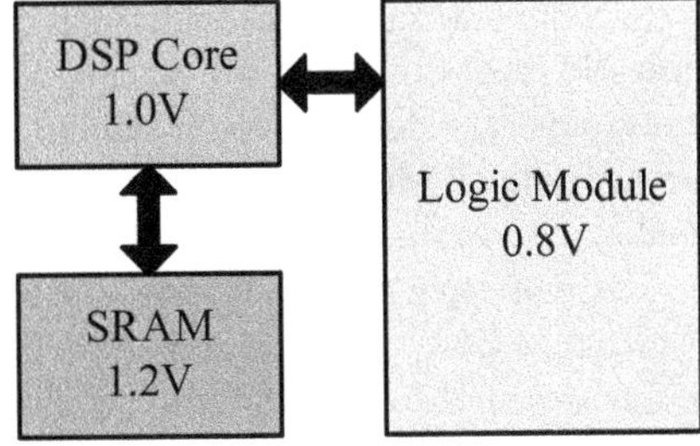

Fig. 6 Multi-voltage architecture of a SOC

power benefits with much smaller area overhead. But it also has some shortcomings, such as increasing latency and design complexity [4].

Using Multiple Supply Voltages

Lowering the supply voltage of selected blocks in the system can also help reduce power consumption significantly. This strategy using multiple supply voltages in a design is often called Multi-V_{DD} [5].

In a Multi-V_{DD} design, the system is subdivided into blocks, which are called voltage regions or power domains, having their own different supply voltages. This method is based on the knowledge that different modules in a modern SOC may have different performance targets and design constraints. Take a high speed SOC as an example, the DSP core runs most fast and needs high supply voltage for its speed determines the system performance. The SRAM, working as the cache, may need even higher supply voltage, because they are usually on the critical paths. The rest of the system maybe need much lower supply voltage to meet performance. Therefore, each block of the system works at the lowest supply voltage consistent with meeting the system timing, and remarkable power saving can now be achieved [1]. Figure 6 gives a framework of a Multi-V_{DD} SOC design.

However, using Multi-V_{DD} adds some complexity to the design. For example, we need a more complex power grid and level shifters (or level converters) on signals running through blocks with different V_{DD}.

Using Multiple Device Threshold

As explained earlier, when process geometries shrunk to sub-100 nm, leakage power became a big problem. Using libraries with multiple threshold voltages has become an efficient strategy bringing leakage reduction without any dynamic power or performance costs [6].

Many libraries today offer two or three versions of their cells: Low threshold, Standard threshold, and High threshold. The high threshold cell features a leakage that may be about one order of magnitude lower than that of the low threshold cell at the expense of some performance loss.

Introducing multi-threshold technique has little impact on the normal design flow. No level shifters or any other special circuits are required when using multiple thresholds. The designers only need to increase the thresholds in timing path which are not critical. Modern EDA tools can usually support multi-threshold design very well.

There are two strategies can be used for multi-threshold design. If minimizing the leakage power is more important than achieving a minimum performance, the design can start from high threshold library followed by swapping in low threshold cells. Otherwise the design process can go the other way around, which means synthesizing with the low threshold library first and then replacing the cells not on the critical path with their high threshold equivalents [1].

Run-Time Power Management

When doing power optimization at design time, we must always ensure that the peak performance that may be required in a system can be achieved. But most digital circuit systems, especially processors or SOC, usually not always work at the heaviest computational load, which means that the activities of the circuits may change greatly over time. For example, in portable applications, the computational tasks to be executed on a DSP/CPU can usually be divided into three categories: compute-intensive tasks, low-speed tasks, and idle-mode operations. Therefore, power optimization for run-time is very attractive. In this section, we will introduce two main low power techniques which exploit the variation in activity or operation mode of a system.

Dynamic Voltage and Frequency Scaling (DVFS)

Fixed supply voltage reducing often conflicts with performance requirement, which is unacceptable. As we know, a system or parts of the system not always work at the highest speed. When the workload decreases, it is possible to lower the clock frequency to reduce power. But by adopting only this method, which is named Dynamic Frequency Scaling (DFS), the total energy consumed for a task will not be saved because every switching occurs still at the high supply voltage [3]. However, if we afford lower supply voltage when reducing the clock frequency, it will lead to both dramatic dynamic power reduction and energy saving.

To achieve substantial energy saving while maintaining the required throughput for peak workload, both the frequency and the supply voltage should be dynamically adjusted based on the current workload of the circuit [7]. This technique is usually called Dynamic Voltage and Frequency Scaling (DVFS).

DVFS brings also some challenges to the system design. Because DVFS is closely relevant with the workload being executed, measuring and predicting the workload accurately, which is usually implemented by software, is extremely important before scaling the voltage and frequency. Any inaccurate estimation will greatly decrease the efficiency of DVFS. What is more, changing the supply voltages introduces some time delay and energy overhead, which can not be ignored. Last but not the least, an additional off chip voltage regulator is needed for the processor or each voltage scaled parts of a SOC, and a voltage level shifter is required whenever a low voltage signal is driving a high voltage receiver.

Dynamic Threshold Scaling (DTS)

Equation (6) shows that a linear change in threshold voltage will bring an exponential change in sub-threshold leakage current which dominates the static power in previous technology nodes. In order to reduce the leakage power, instead of adjusting the supply voltage and frequency in DVFS, the Dynamic Threshold Scaling (DTS) controls the threshold voltage according to the current workload, by controlling the body bias. This technique is especially attractive in light of the increasing impact of leakage power [8].

Every MOS transistor has a fourth terminal (the substrate), which can be used to scaling the threshold voltage by reverse or forward biasing. For high speed computation tasks, the threshold voltage should be adjusted to the minimal value; for the background tasks or high-latency tasks, which can be executed at a reduced frequency, the threshold can be changed to a higher value to reduce leakage; when the system is in the standby mode, the threshold should be set to the highest value [9].

It is alluring that the DTS technique does not change the circuit topology, does not need any voltage level shifter, and comes without any performance decrease. Generally speaking, DTS is easer to be implemented than DVFS. But the drawback is that it is required to control all the four terminals of both NMOS and PMOS transistors independently, which can only be realized in the triple-well technology. Though it is proved in [8] that DTS strategy is effective in reducing leakage power for 70 nm process, its effectiveness is rapidly decreasing with the scaling of the technology below 100 nm because the range of threshold scaling by dynamic body biasing is limited.

Operand Isolation

Operand isolation is often used to save dynamic power dissipation in data-path by reducing the unnecessary switching activity. Today's digital circuit designs usually contain many data-path modules that only occasionally perform useful computations but spend a remarkable amount of time in idle states. However, switching activity at their inputs in their idle states leads to redundant computations which are not used by the downstream circuits. The power dissipated by the redundant computation activity is just a great waste!

Operand isolation was first introduced in IBM PowerPC 4xx-based controllers [10]. The main idea of this method is to identify the redundant operations and, utilizing special isolation circuitry, prevent switching activity from propagating into a data-path module whenever it is about to perform a redundant operation. Therefore, the transition activity of the module is reduced significantly, resulting in lower power dissipation [11, 12].

Operand isolation technique is very useful in applications which employ complicated combinational modules, such as ALU, long word adders and/or multipliers, and hierarchical combinational cells. DSP/CPU is a good example of the applications. In a particular clock cycle, usually only one of the data-path modules of the DSP/CPU executes the useful computation, other modules are just idle. Therefore, operand isolation can be used to control the redundant computation activity and reduce dynamic power significantly.

Operand isolation circuits can be easily inserted by using modern EDA tools. However, one important issue which should not be ignored is the leakage power. We should ensure that the isolation circuitry is designed and used in a way that the isolated module consumes minimal leakage power as well.

Reducing the Power in Standby Mode

In many systems, especially mobile application systems, standby (or sleep) mode takes great part of the operation time. Power management for this mode is critical for the battery life. In standby mode, total or parts of the system have no useful computational tasks to execute, and the dynamic power and standby power are hoped to be zero or very low. Many techniques are proposed to control the power dissipation in during period of inactivity, and we will introduce some main of them below.

Clock Gating

In the digital circuits, the clock signal is usually highly loaded. The power consumed by clock network (often a clock tree) contributes a great part of the total dynamic power [13]. Especially in standby mode, the clock switching is the main source of the dynamic energy consumption. In addition, when the registers receiving the clock maintain the same value, they will still dissipate some dynamic power.

The clock gating technique stops the clock signal fed into the idle modules, so it can significantly save the dynamic power not only from driving the clock tree but also from the unnecessary gate activity. Clock gating can be used in parts of a system or over the total circuit.

Nowadays, most of the libraries include specific clock gating cells. Meanwhile, modern EDA tools can automatically identify circuits where clock gating cells can be inserted without affecting the function of the logic. What is more, clock gating has many other advantages, such as low hardware overhead and little performance penalty. All of these make clock gating a simple, reliable and most widely used method of reducing power consumption.

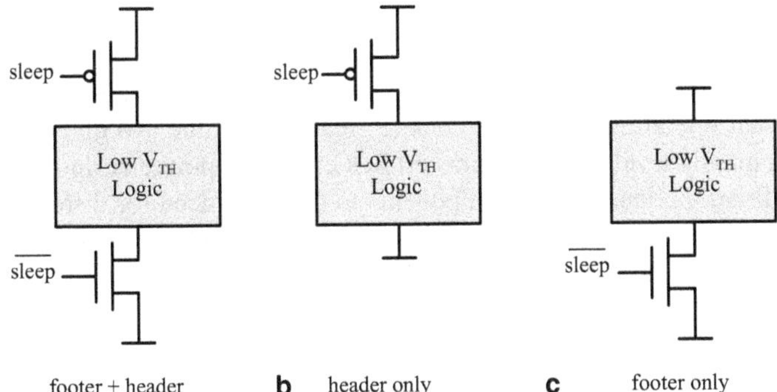

Fig. 7 Switches in power gating design

Power Gating

After adopting clock-gating method, the power dissipation in standby mode is mainly from leakage, which may decide the battery life of portable devices and should not be ignored in today's sub-100 nm technology. The power gating technique reduces both dynamic power and leakage power by cutting off the power supply of the idle blocks of a design, which is an important advantage over clock gating. For the systems or sub-systems in standby mode for long period of time, this technique is extremely attractive [14].

However, power gating is much more difficult to implement than clock gating and brings higher costs. One of the main challenges for power gating is the design of the proper power switching fabric. The common approach is to use the large size sleep transistors, either PMOS or NMOS devices of high threshold voltage for low leakage, to switch off the power supply rails when the circuit is in standby mode. The PMOS sleep transistor, often called "header switch", controls V_{dd} supply; and the NMOS sleep transistor, often named "footer switch", is used to switch GND supply [15]. In real design, either header switch or footer switch, or both, can be used to implement the power switching fabric. However, in designs of sub-90 nm process, either PMOS or NMOS switch is only used because of the area penalty of the large sleep transistors and the constraint of the sub-1V power supply. Figure 7 illustrates the switches in power gating design. The sleep transistors must be carefully optimized so that the benefits of leakage power saving overwhelms the area and power penalties from sleep transistors [15].

Compared to clock gating, another shortcoming of power gating is that it will take some time for a system or block to enter and leave the power gated leakage saving mode, which introduces additional delays and power penalty. What's more, the power gating method affects the inter-block communication of the system, and reduces the noise immunity if care is not exercised when designing the sleep transistors.

Other challenges in power gating designs include: design of isolation cells between sleep blocks and active blocks, design of the power gating controller, selection and use of retention registers for data and states, and so on.

Dynamic Power Management (DPM)

Processors, such as DSPs and CPUs, are special digital circuits with good flexibility, and their operating mode may change greatly over time in a given application.

DPM, mainly used in processors and SOC, is a low power design methodology, which utilizes the least active components to meet the need for different computation tasks, by dynamically configure the system parameters. DMP encompasses a set of techniques to selectively shut off the components in idle mode, achieving a highly energy-efficient system [16].

DMP needs the support of hardware and software. Software predicts whether a device can go to sleep long enough to save energy, mainly based on two kinds of policies, which are predictive and stochastic [17]. Hardware support mainly includes operating mode management and some basic low power method, such as clock gating, DVFS, power gating.

The basic idea of the operating mode management is to cutting off the clock signal or power supply to the circuits when they are in sleep mode. Besides active mode, the most common low power operating modes (usually called sleep modes) in processors are: idle mode, standby mode and power down mode. In sleep modes, the frequency or supply voltage can be reduced, and the modules that are not invoked are hung to save power dissipation.

2 Application Specific Instruction-Set Processors

2.1 Introduction

The embedded system face many design constraints, such as performance, power consumption, chip cost and time to market. ASICs (Application Specific Integrated Circuits) are designed for a specific application. The chip area, power consumption and performance can be optimized easily. However, the ASIC design requires an extensive verification time. GPPs (General programmable processors) make a short design time possible due to the high flexibility, but they may not meet the challenges of the performance and power consumption. ASIPs (Application Specific Instruction set Processors) provide a compromise between ASICs and GPPs shown in Fig. 8.

An ASIP is a processor designed for a specific application or application domain, typically consisting of a programmable base processor, various customized instruction set extensions and customized hardware accelerators. The structure of a typical ASIP is shown in Fig. 9. Thus, the instruction sets of ASIPs are tailored to the applications. Through architectural customizations ASIPs explore potential power optimization.

Due to the well-defined instruction set, ASIPs offer high programmability, which makes the system using ASIPs can be software upgradeable. At the same time, the customized instruction set extensions and hardware accelerators make the implementation of ASIPs to meet the challenges of high computing performance and

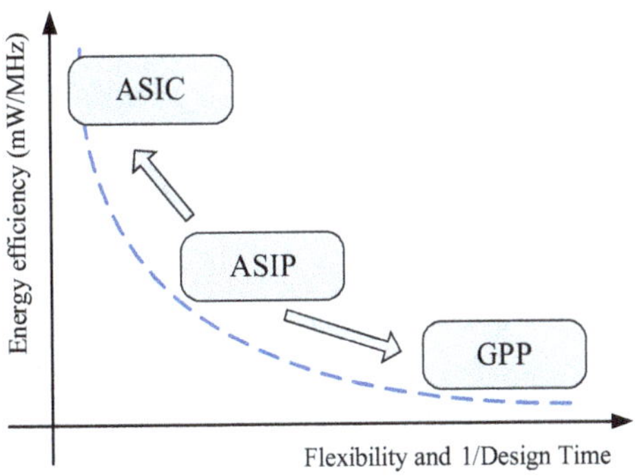

Fig. 8 Comparing of ASIC, ASIP and GPP

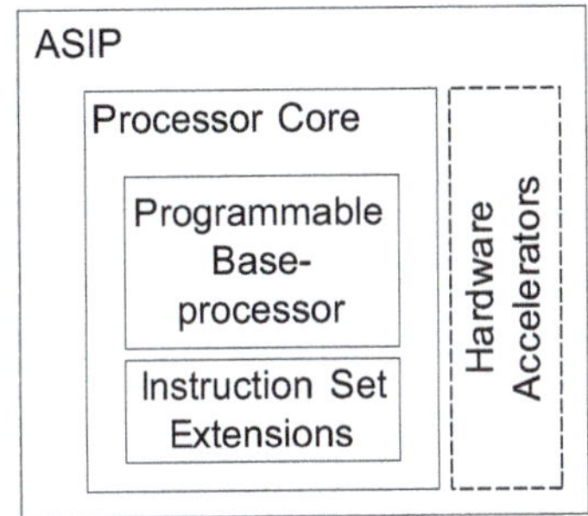

Fig. 9 Structure of the ASIP

low power consumption. ASIPs can be efficiently used in many embedded systems, such as digital signal processing, video surveillance, automotive control and individual health systems.

An ASIP exploits special characteristics of applications to meet the computing performance, cost and power requirements. Typically there are five steps in the ASIP design flow, which are respectively application analysis, architecture design space exploration, instruction set generation, code synthesis and hardware synthesis [18]. A simplified block diagram is given in Fig. 10.

Application analysis takes an application or a set of applications written in high level languages as input and identifies the hot spots, which are the most time consuming or the most energy consuming parts of the applications. The results will be carefully optimized in the following phases.

In architecture design phase, according to the results of application analysis and the given design constraints, a possible architectures design space is explored. To choose the proper architecture, performance and power of the possible architectures are estimated.

Next, the instruction set of the selected architecture is generated and enhanced by adding a set of customized instructions using the results of application analysis.

In code synthesis phase, a compiler will be generated for the new instruction set architecture and the executable code for the application will be emitted. The designers can verify whether the design meets the given constraints, which is important and should be iterated fast to provide design feedback.

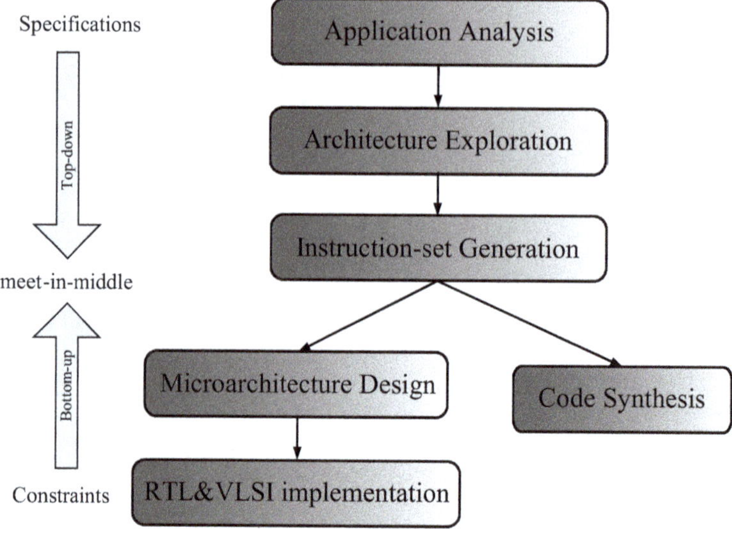

Fig. 10 ASIP design flow diagram

When the design meets all the constraints, the microarchitecture of ASIP is implemented. The architecture design includes selections of processor structure, available modules, and the interconnection between modules. Microarchitecture design will further map each instruction function into hardware module. Then follow the RTL coding and VLSI implementation.

The crucial work of the ASIP design is usually carried out manually, such as code segment identification, extensible instruction generation and processor evaluation. While ASIP design environments are getting more and more effective, problems do exist, for example, how to get the optimization of ASIP architecture and how to realize an efficient software development environment [19–22].

2.2　Design of Low-Power ASIP

Application Analysis

Application analysis is the first step of ASIP design, which usually means design space exploration at system level. It takes an application or a list of applications written in high level languages as input and identifies the hot spots, which are the most time consuming or the most energy consuming parts of the applications, through analyzing the source code structure and estimating the execution time, memory cost of the source code and so on. The analysis results are the code analysis and execution statistics, which are the key to eliminate the critical path of the application, thus increasing the performance, reducing the runtime and power consumption.

Code profiling is widely used for application analysis [23]. To ensure the effect of the analysis result and the flexibility of the ASIP, source code should be well selected. Profiling exposes the behavior of the program entities, such as functions, loops and subroutines. Most importantly, it also gives the cost of the design, including execution time, the code structure, run-time cost and hardware usage. Execution time is usually presented in the form of the number of operations, which requires to be translated into the number of clock cycles according to the selected architecture. Particularly, the direct intention of code profiling is to expose three kinds of application costs, which are execution cost, memory cost, and hidden cost.

1. The execution cost depends on application computing task directly. While profiling the high level source code, some costs may not be located. Designers can introduce reference processor architecture to estimate the execution overhead. Execution cost includes the followings:
   - Inputs and outputs synchronization cost
   - Datapath cost: word width and rate of the runtime execution
   - Memory usage
   - Program flow control cost
   - Cost to offer parallel task

The profiling results can locate the performance critical path, such as the most time consuming functions, the most frequently using operations and the largest memory size used during running time. Designers will follow those guides during architecture and instruction set exploration.

2. The memory cost is the totally number of storage bits used to temporarily store data and control information. The memory cost analysis includes the total memory access counts (read and write operations) and particular accessing behavior while running a task. The profiling results may include the following factors:
   - Total number of memory accesses
   - Percentage of Load/Store operations
   - Number of addressable words and word width per word
   - Hierarchy: register files, on-chip memory, off-chip memory
   - Access timing constraint: cycle counts to read/write memory
   - Memory organization: single and dual ports memory, synchronous and asynchronous output

Memory accesses are expensive in terms of power consumption, due to the switching on heavily loaded internal bit- and word-lines. The average power per memory access increases with memory total size. The main purpose of memory cost optimization is the reduction of memory accesses and memory size in order to reduce the area and to increase the computational throughput of an implementation.

3. The hidden cost is hard to be recognized because profiling is traditionally hardware independent. Template architecture can be used to locate the hidden cost. The hidden costs could be from:
   - Cost for the processor stalling during subroutine calling and returning
   - Cost for system hardware initialization
   - Cost for thread and interrupt handling
   - Data hazards, control hazards, and structural hazards
   - Cost for synchronization when data transmit across asynchronous clock domains

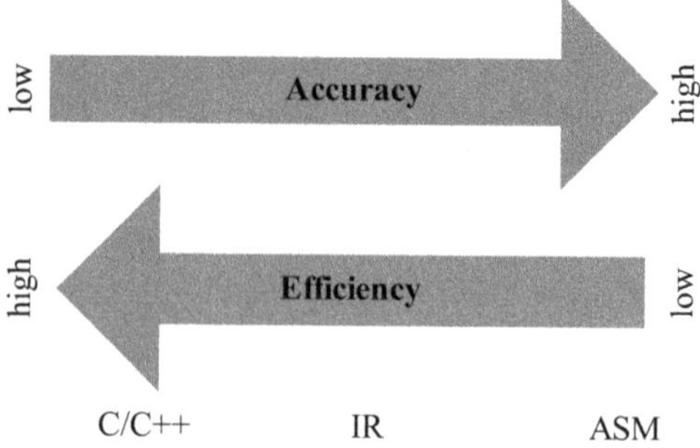

Fig. 11 Comparing of three kinds of code profiling

Profiling can be performed on source code level or assembly level. Source code level profiling is fast, but has a gap to the instructions executed on the instruction set architecture (ISA). A line of C code may consist of several operations and can be mapped to several instructions. Some assembly level operations are invisible to the source level profiler, such as address computations, memory accesses and type casts. Also a C compiler optimizes the source code using various compiler techniques, which will be ignored by the source code level profiler. The source code level profiling is not accurate to design the ISA [23].

Assembly level profiling is accurate to design the ISA based on a detailed processor architecture model. We can use an available architecture and relative assembly fitting the application as a reference for assembly level profiling. The gap between the expected instruction set and the reference instruction set can be identified to get the new instruction set. In addition, we can use intermediate representation (IR) based on open compiler framework. The high level source code is lowered to optimized and executable IR. All the operations, such as arithmetic, logical, type casts, address computations and memory accesses, are visible in the IR. The profiling based on the IR is accurate and efficient enough (see Fig. 11), at the same time no available processor architecture is needed.

According to the analysis results and the given design constraints, the design may be implemented using a scalar ASIP. To achieve a higher design goal, the ASIP may be implemented as a data or instruction level parallel architecture, for example, a SIMD or a VLIW implementation. In this section, we just discuss the scalar ASIP.

Architecture Exploration

The first step of architecture exploration is partitioning the application into blocks with reasonable size, and mapping each of these blocks to either software implementation, to a tightly coupled accelerator or to dedicated circuits. Given an application without high computing task, both hardware and software implementations are reasonable. Then, other constraints have to be further considered, such as energy-efficiency and flexibility. The partitioning task is difficulty, because high level profiling results tend to be inaccurate. Designers may have a coarse idea of the ASIP instruction set and architecture, however, the details of the target implementation is still unknown. Facing a large design space, several issues should be paid attention to locate the target architecture subset.

4. Datapath

The datapath represents the computing elements of a processor core. It gets operands from register file and sends the results back to the memory subsystem. The datapath function covers basic arithmetic, logic computing and some application-specific operations. The following parameters can be identified for the datapath:

- Granularity: bit serial, byte serial, word serial and word width
- Number of operands per instruction
- Arithmetic: fixed point, block floating point or floating point
- Functions: basic integer arithmetic and logic, multiplier and accumulation, dedicated algorithm function (divide and counting)
- Hardware multiplexing: mapping multi-operation into one hardware kernel
- Number of execution cycles
- Accelerator: with optimized dedicated hardware structures for extremely high performance or extremely high energy-efficiency requirement.

5. Memory subsystem

The memory is a hardware storage component where data can be stored and retrieved to and from a physical position inside the module. Each position is specified by a unique address, and the data access (storing and retrieving) is to point to the corresponding address with associated access control (read or write). A memory subsystem includes physical storage elements, address generators, memory peripheral circuits, and memory buses. The following list of characteristics determines the memory subsystems for an ASIP.

   Hicrarchy: register files, on-chip memory and off-chip memory
- Addressing bus: uniform addressing or individual addressing bus
- Access time: number of clock cycles per read/write operation
- Memory organization: single or dual ported memory, output registered or not
- Capacity: number of words and word width

6. Control path

The control path manages the program flow and supplies the control signals for data path. The main task of a typical control path is to handle the pipeline hazards. Pipeline hazards are due to the dependence between instructions that are so close that the overlapping execution in pipeline leads to a different access sequence of resources or data. Functions of the control path include:

- Control for program flow, such as PC FSM, hardware loop control, stack and jump
- Interrupt handling factors: number of interrupts supported, interrupt priority definition, interrupt latency, return from an interrupt

- Control signals to datapath, such as address generators, ALU, MAC, register files
- Control signals to solve pipeline hazards, such as data hazards (read after write, write after write and write after read), control hazards and structural hazards.

7. I/O access mode

Data transmission between ASIP core and other external components such as processors, devices or analog IP, can both affect the system overall performance and power consumptions. Different I/O access mode can be used with respect to data transmission characteristics such as system throughput, sample rate, and data frame size.

- Memory-based access: The input and output buffer is used as a shared memory. Data can be accessed by ASIP internal memory bus. This method is suitable for high data rates transmission.
- Register-based access: The data buffer is used as a part of internal registers and can be accessed by register bus. This method is typically suitable for a smaller data transmission. Additional instruction will be created to support the access operation to the input/output data registers.

All top-level features for ASIP design space have been addressed with the goal to provide well-defined degrees of freedom for ASIP designer. Therefore, we can design an ASIP with its own features for the target application, for example, to enhance the execution performance, or to minimize the power consumption for certain algorithms. The microarchitecture implementation details for the ASIP design will be discussed later.

Instruction Set Generation

Since the processor architecture has been proposed based on requirements of performance, flexibility, and cost for target application, the instruction set design has started. The ASIP instruction set design can be divided into design of general RISC instructions subset and design of extended instructions for customized functions. The RISC instructions subset handles the general arithmetic and control operations, which is used as a basic instruction template.

The basic instruction subset can be optimized by adding instruction set extensions performing several frequently used operations in parallel to reduce the run-time and the power consumption. Code segment identification [24] tries to find small and reusable DAG patterns from the application hot spots and the instructions are added during the design of the processor core.

Computationally intensive or high memory bandwidth algorithm may be mapped to dedicated hardware accelerator when the processor core cannot offer enough computing performance. Code segment identification tries to find large, regular and

Table 1 Basic Instruction Set for Profiling

Type of Instruction	Instruction mnemonic	Description
Load/Store	RDIO	Read I/O data
	WRIO	Write I/O data
	RDM	Read data memory
	WRM	Write data memory
Arithmetic	ABS	Absolute value
	ADD/ADDI	Addition
	MOV/MOVI	Data move
	MULU/MULS	Signed/unsigned multiplication
	SHL/SHLI	Arithmetic/logic shift left
	SRA/SRAI	Arithmetic shift right
	SHL/SHLI	Arithmetic/logic shift left
Logic	AND/ANDI	Bitwise AND
	OR/ORI	Bitwise OR
	SRL/SRLI	Logic shift right
	XOR/XORI	Bitwise XOR
Control	CMP/CMPI	Compare/set status
	BRA	Unconditional branch
	BSR	Branch to subroutine
	BEQ/BNE	Branch if equal/not equal
	BLT/BLE	Branch if less than/less or equal
	BGT/BGE	Branch if greater than/greater or equal
	END	Exception/transition to idle mode
	RTS	Return from subroutine

stable sub-graphs from the application hot spots, for example a FFT (Fast Fourier Transform) unit, and the instructions are added after the design of the processor core.

The output of the instruction set design is an assembly instruction set manual and requirements for processor microarchitecture design, which describes the function specification, and assembly code, and the binary coding for each instruction.

Basic Instructions Set Template

Basic instruction set template can be found from publications and open sources. Generally, basic instruction template contains three kinds of instructions: basic data move and load/store instructions, arithmetic and logic operation instructions, and instructions for program flow control.

A basic instructions template with two operands is described in Table 1.

To improve the processor performance, the two operands often directly come from register files or immediate data carried in instruction code. The data load/store instructions load operands from memory to register, and store computing results back to the data memory. The explicit I/O instructions have been provided in order to profile the input/output behavior of certain applications. Basic arithmetic and logic instructions include all arithmetic, logic, and shift operations executed in the datapath. Since functions like division and modulo operations are usually expensive due to hardware cost, energy or latency, they have been removed from the template. The program flow control instructions perform branch, jump, loops, subroutine call and return function, including testing conditions and jumping to the exact position of program memory.

Some seldom used RISC instructions could be removed from basic template. For example, an ASIP do not often uses the unsigned arithmetic and string operations instructions. These instructions should be removed for hardware gates saving. Contrastively, some frequently used specific operations are not supported by general RISC instructions, such as an operation counting leading identical bits from MSB. This kind of operations could be mapped into additional instruction to significantly improve the performance. Besides, if several instructions appear together mostly for certain function, combining these instructions into a new one can significantly improve the computing efficiency and program memory capacities.

Instruction Set Extension

One main optimizing approach in ASIP design is to extend the instruction set by adding customized instructions. Instruction Set Extension (ISE) relies on identification. A typical identification algorithm looks into the Data Flow Graphs (DFG) of each hot spot and attempts to merge several operations into one instruction. The results of the identification are a set of ISEs represented as Directed Acyclic Graphs (DAGs). Finally, the DAGs are converted to customized instructions and mapped to data path inside the hardware model.

Every customized instruction gives a better code quality, such as code size reduction and runtime reduction. Runtime reduction in general means lower power consumption for embedded system. Customized instructions are designed in the execution stage of the base processor. The amount of computation performed in the instruction should be limited to avoid the clock period violation. If violation is inescapable, split the customized instruction into multiple instructions, or multi-cycle the execution of the instruction. If the execution of the instructions requires multicycle, an appropriate scheduling should be taken into account.

For example, a Maximize instruction of two operands frequently used in the application, using the base processor ISA we must move one operand to the destination register and compare the two operands, and then we move the other operand to the destination register conditionally based on the ALU flag, or use branch instruction. If a new instruction executing the same operation in one cycle added, it can replace the sequence of the base processor instructions, which can significantly reduce the code size and the run time, thus improve the energy efficiency.

For an application with lots of memory access operations, instructions including memory access could be added to reduce the number of memory load and store. Power consumption produced by the fetching and decoding of the instructions and the relatively register accessing can be avoided.

Instruction set extension could maximize performance while meeting chip area and power consumption requirement for certain applications, but must base on the base processor.

There are two important features for instruction set extension: the extra coding space and following the coding protocol, so that the new instruction can share the processor data path and control path. The extra coding space should be considered for future extension when designing the instructions set template. The coding protocol exposes ways to reuse the processor hardware, such as operand fetch and results write back. Additional decoding logic and dedicated hardware will support the extended instruction to be executed as normal assembly instruction.

Hardware Accelerator

If instruction set extension can not satisfy the computational performance or power consumption requirement, hardware accelerator can be implemented. A hardware accelerator is a dedicated hardware module to enhance the performance or functionality.

Besides powerful computing ability, hardware accelerator can reduce the run time, code size and power consumption at the cost of additional area consumption. But it may decrease the flexibility of the design, because accelerator usually performs limited computational tasks. Also, the hardware accelerator may not be used any more, if the algorithm changes. So hardware accelerator can be added only when the processor core can not offer enough performance to certain high computing demand functions, such as FFT, or certain functions missing during the instruction set design, such as integer division.

The core should provide a well-defined protocol that can be followed to add an accelerator. The interface and control mechanism should be part of the protocol content. The interface passes data between the ASIP core and the accelerators and can be realized using special registers or general purpose registers. The control mechanism starts the accelerator and synchronizes the results with the program flow and can be realized using configurable register or special instructions. Usually hardware accelerators need more execute cycles than the base processor instructions and the customized instructions. The designer should be responsible for the synchronization between the hardware accelerators and the base processor. When adding or removing an accelerator, the design of the core should not be modified to minimize the risk of redesign.

The accelerator can be implemented using one instruction or multiple instructions. The instructions indicate configuration information such as the control signals and operands. One-instruction accelerator can be used to simple functions with fixed cycle cost, such as integer divider, operations obtaining square root, logarithm and reciprocal. Multi-instruction accelerator can be used for relatively regular functions building upon its simple instruction flow. Multi-instruction accelerator provides more flexibility by different combinations to realize more corresponding

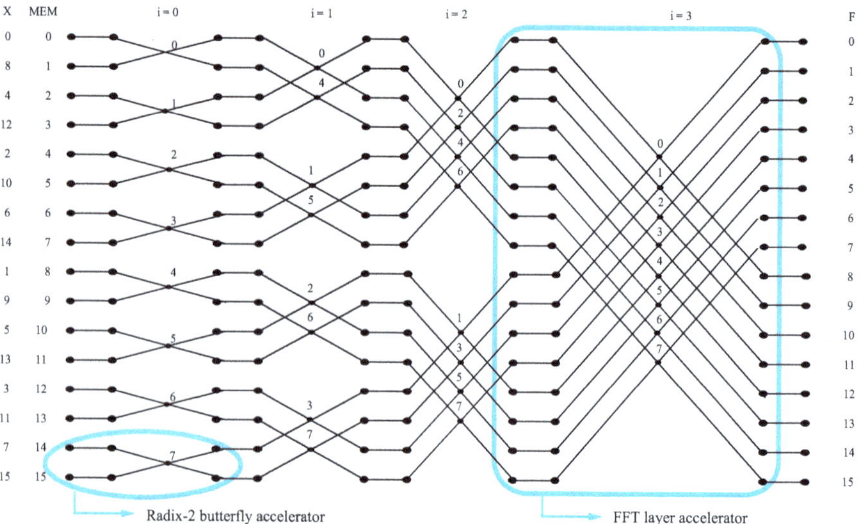

Fig. 12 Fine and coarse acceleration strategies for a 16-point FFT task

functions [23]. For example, in a Multi-instruction FFT accelerator, each layer FFT can be performed by one instruction. The processor core arranges the instructions flow for FFT. Another fine granularity of acceleration is done in such way that combining each radix-2 butterfly operation into dedicated hardware, as shown in Fig. 12. Computing a radix-2 task generally consists of about 10 operations, including totally 5 memory accesses (two data load, one coefficient load, and two data store), two data additions and a complex data multiplication. Thus, there is a speedup about $5 \sim 10$ times if the radix-2 butterfly can be executed by accelerator in 1 or 2 cycles.

Also, a hardware accelerator could be a slave processor. The master processor just sends a task code to the slave processor to initialize the task. Since the slave processor is programmable, it can execute many complicated tasks at the cost of increased chip area. The master and slave processors run in parallel, thus the throughput of the system is improved obviously. Proper partition of the application to processors can make the system to work efficiently [23].

Code Synthesis

Software development tools like compiler, assembler, linker and simulator are required in ASIP design. In code synthesis phase, the tools will be generated for the new instruction set architecture and the executable code for the application will be emitted. To make the ASIP design feasible, the software development environment should be implemented in a short time.

Optimized hardware implementation that matches the applications is the key in ASIP design. Software development tools can be used to evaluate the design and provide design feedback, thus the designers can verify whether the design meets the given constraints and customize the processor architecture effectively.

Software implementation efficiently using the hardware resource can reduce the power consumption. Compilers tailored to the ASIPs can efficiently map program code of the application to the customized instructions or customized accelerators, sometimes, relying on the inline assembly or intrinsic functions. At the same time, compilers usually use various compiler optimization techniques to yield high quality object code, and optimization reducing the run time can also reduce the power consumption. Low-power compiler optimization strategy can reduce the power consumption obviously. Energy cost driven code generator selects instructions based on the Instruction Level power model of the target machine. Instructions reordering can reduce switching between consecutive instructions, thus reducing the power [25].

Optimizing algorithms can also reduce the power consumption [26]. Usually, algorithms requiring more runtime and operations consume more power. Complex algorithms need more operations. Irregular algorithms need more program control flow instructions, thus resulting in branch penalty for taken branches. Memory accessing usually consumes more energy, thus algorithms should reduce the number of memory access and memory size required to improve energy-efficiency. For example, realizing a 2N-point FFT of a real sequence uses a single N-point FFT. The runtime, the amount of computation and the number of memory access will all decreased, thus power consumption decreased. Power consumed by iterative algorithms increases with the iterative times, therefore proper iterative times and precision of the results can reduce the power dissipation. Algorithm implementation should follow the advices of the compiler user's guide to efficiently use the hardware resource. Appropriate algorithm statements can help the compiler to recognize the variable hardware used. Using inline assembly or intrinsic functions can help the compiler to map the algorithm directly into the hardware.

Only when the ASIP architecture and the ASIP software are optimized jointly, we can get a power optimization ASIP design.

Microarchitecture Design

In general, microarchitecture design is the detailed hardware implementation for each instruction under performance and power budget. It involves pipeline stages scheduling, partitioning instruction function into several micro operations, and mapping each operation to specific hardware modules, hardware multiplexing and minimization. At microarchitecture level, an ASIP core includes data path (ALU, Register File and MAC) and control path (Pipelines, Program Counter, Instruction Decoder, address generator, pipeline control). Essential hardware components of the microarchitecture will be discussed in this section.

Cycles

Instructions	1	2	3	4	5	6	7	8	9
1	IF	ID	OF	EX	WB				
2		IF	ID	OF	EX	WB			
3			IF	ID	OF	EX	WB		
4				IF	ID	OF	EX	WB	
5					IF	ID	OF	EX	WB
6						IF	ID	OF	EX
7							IF	ID	OF
8								IF	ID
9									IF

Fig. 13 Pipeline organization of a typical RISC

Pipelines

The concept of pipeline is introducing additional storage elements to subdivide groups of functional units into different stages for instruction execution. The effect of pipeline is to increase the computational throughput by working at higher clock frequency. This is especially attractive, if the same processing task has to be performed for a series of input data. Pipeline is also a technique to utilize functional units more efficiently, because the functions are partitioned into micro operations that can be executed in parallel, which is just like the concept of an industrial assembly line. A typical pipeline organization of a RISC processor uses the pipeline stages like Instruction Fetch, Instruction Decode, Operand Fetch, Execute, and Write Back. Functions in different pipeline stage that are executing in parallel are illustrated in Fig. 13. There are five different hardware modules that running in parallel from 5 to 9 cycles. Given the sixth cycle as an example, there are five instructions operation simultaneity mapped to the pipeline: the sixth instruction is fetched, the fifth instruction is decoded, operands of the fourth instruction are fetched, the third instruction is executed, and the second instruction's results are written back.

Assume an instruction takes time T to finish the execution without pipeline. When partitioning the execution into n steps, then each execution stage takes time T/n. By queuing these steps into n independent hardware pipeline stages, the total execution time will still be time T. But the system clock frequency will increase n times. Actually, the speedup factor will reduce due to the imbalanced task partitioning and pipeline hazards. By experience, a five-step pipeline achieves 2 to 4 times speedup.

An essential principle of pipeline design is to balance each pipeline stage execution time. When a critical timing path exists in a certain pipeline stage and the timing of the next or previous stage is not that critical, the pipeline timing can be

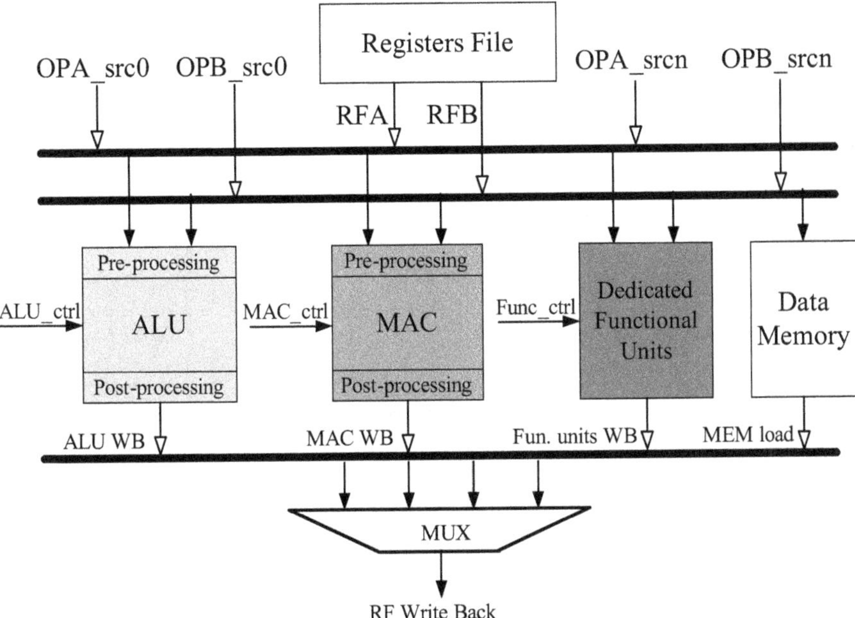

Fig. 14 DSP datapath including ALU, MAC, and function units

balanced by moving parts of operations to the neighboring pipeline stage. Thus the critical path timing will be relaxed. Using an ALU stage with critical timing path as an example, designer can either move operands pre-processing to OF pipeline stage or to move computation results post-processing to the WB pipeline stage.

ALU

Computing units usually consist of ALU, MAC and Dedicated functional module, as shown in Fig. 14. They get operands from a RF or immediate data directly from instruction, and send computation results back to a RF. The basic RISC instructions including arithmetic and shift instructions are usually executed in ALU, and the execution cost for each ALU instruction is one clock cycle. The MAC unit is a hardware module including a multiplier and an accumulator. Thus, the double-precision computation task can be executed by MAC. An instruction level accelerator can be added into the datapath to support specific functions. These functional units often take one or two clock cycles to finish task in order to avoid introducing any pipeline hazards. The kernel components in datapath are full adder or a multiplier that performs the essential function of a module. In order to minimize the silicon cost and power, it is very important to multiplex the kernel components as much as possible.

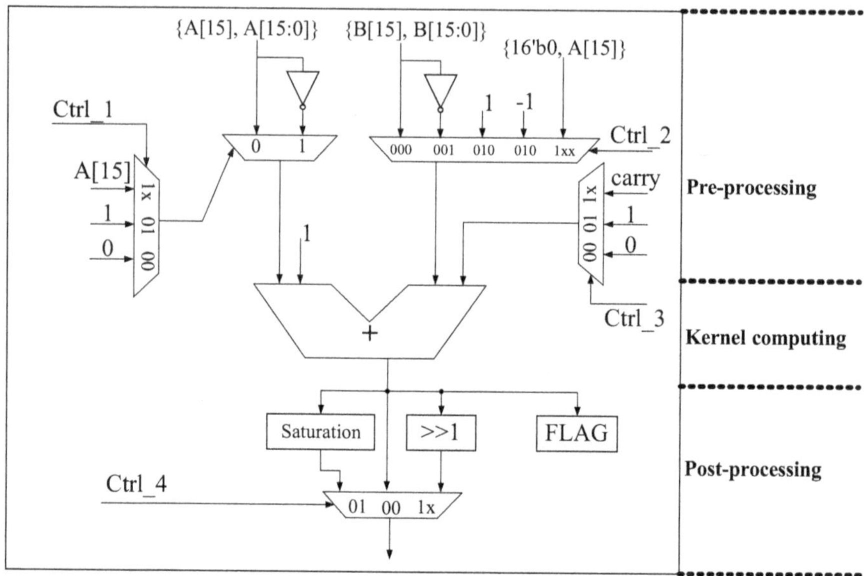

Fig. 15 ALU module with hardware multiplexing

An ALU with detailed hardware description [23] is used as an example to show how to integrate all arithmetic function with hardware multiplexing, illustrated in Fig. 15. The pre-processing of operands includes guard extension, inversion, and carry-in selected. Then the kernel components execute the computation with real arriving operands. The post-processing includes saturation on results and the flag computing. The ALU control signals for both pre and post-processing come from the instruction decoder. The extra hardware cost introduced by multiplexing is ignorable compared to the kernel component cost.

Different arithmetic instruction can be executed on the ALU with respect to different control signals combination. For the operation of two operands addition with saturation, the four groups of control signals should be assigned like this: ctrl_1 = 2'b00; ctrl_2 = 3'b000; ctrl_3 = 2'b00; ctrl_4 = 2'b01. For the operation of two operands subtraction without saturation, the four groups of control signals should be assigned like this: ctrl_1 = 2'b00; ctrl_2 = 3'b001; ctrl_3 = 2'b01; ctrl_4 = 2'b00. Other complex instructions using full adders as kernel devices can also be supported by the ALU. An absolute operations can be executed in such way that: the four groups of control signals should be assigned like this: ctrl_1 = 2'b1x; ctrl_2 = 3'b1xx; ctrl_3 = 2'b00; ctrl_4 = 2'b00.

All we discuss above is only about the arithmetic computation feature in ALU. The other two features of ALU are shift and logic operation. The shift and rotation supply functions like data scaling, bit test, software division and different kinds of coding. Logic operations can be used for normal AND, OR, XOR, INV functions. The idea of hardware multiplexing can also apply to shift and logic hardware module design with the purpose to minimize silicon cost and chip power.

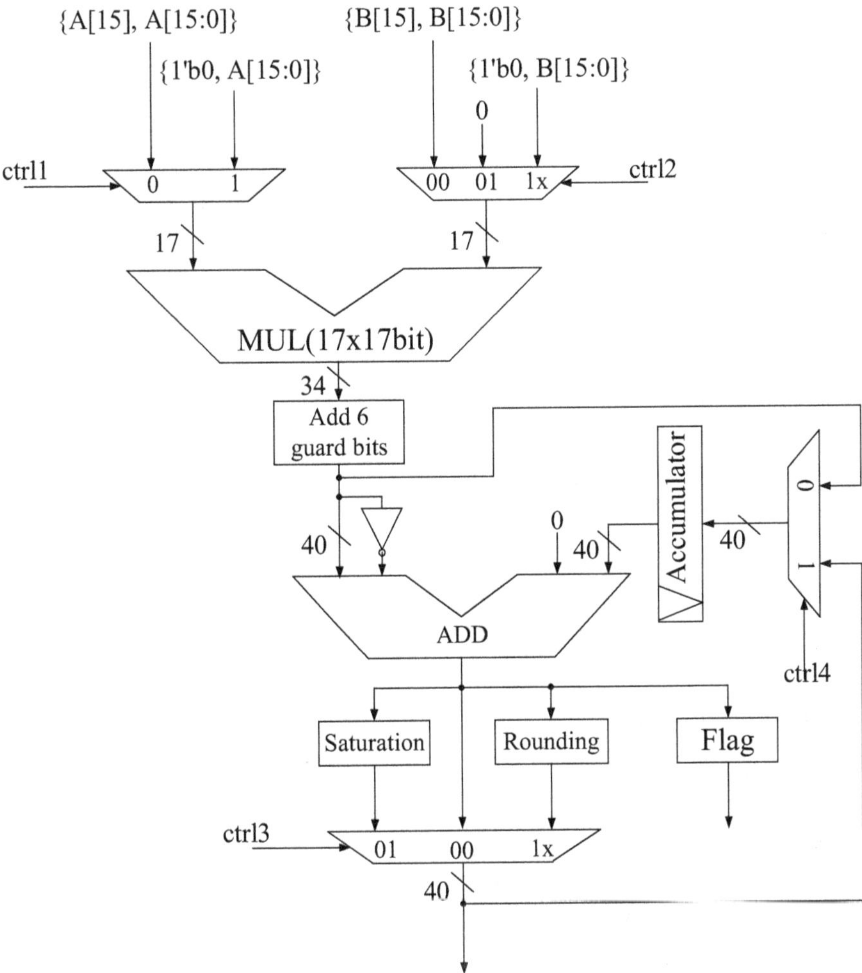

Fig. 16 MAC units

MAC

The multiplication and accumulation are probably the most important operations
in DSP kernel algorithms, such as FFT, DCT, and convolution. A dedicated MAC
hardware unit can achieve a great computing acceleration for DSP application. The
kernel of a MAC unit consists of a multiplier and an accumulator. The basic multiplier is a signed integer multiplier primitive. For a 16-bit data width processor,
the multiplier is usually implemented using two 17-bit width operands as input and
(N+34) bit width as multiplication results, where the N stands for the number of the
guard bits. The accumulator is a two's complement full adder with double-precision
to support iterative computing. Some extra logic is added to the MAC kernel components to support general arithmetic and logic function with high precision. A simplified MAC hardware is illustrated in Fig. 16. Multiplexing hardware is introduced

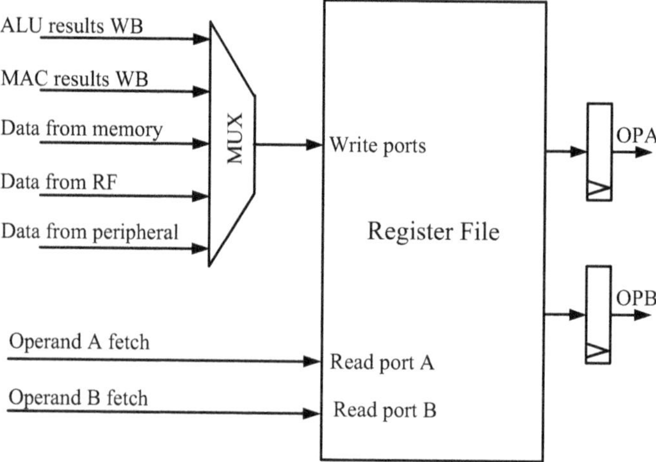

Fig. 17 Simple example of a RF

to support various MUL and MAC operations. The accumulator registers store the multiplication and accumulation intermediate results during iterative computing.

Various instructions using a multiplier as kernel device can be executed on the MAC units. An unsigned multiplication can be executed from the signed two's complement multiplier in such way that adding a 1'b0 as the sign bit on the left side of the operand. For a typical MAC operation of two signed operands, the four groups of control signals should be assigned like this: ctrl1 = 1'b0, ctrl2 = 2'b00, ctrl3 = 00, ctrl4 = 1. According to different combination of control signals, operation like signed/unsigned multiplication, multiplication and subtraction with double-precision can be supported by the MAC.

Register File

Figure 17 illustrates a simple example of register file module. A general register file (RF) consists of a set of registers used as the first level of computing storage elements. The size of a RF is specified during the instruction set design. Today DSP processor often has a RF size of 16 to 64 registers. Too many register in a RF require expensive silicon cost. Too small a RF will introduce frequently data swapping between the RF and data memory.

A typical RF has write port and read port, including address bus, data bus, read/ write enable signals and clocks. The RF should be able to supply two operands fetching simultaneously due to the general arithmetic instruction often includes two operands. The RF supplies at least one write port for data loading. The data sources come from various components, including computation results from ALU and MAC, and memory data by load operation, and RF itself by data move operation, and data from system peripheral. The RF may has two or there write ports to support

Table 2 PC FSM state definition

Processor state	Next Gen.
System reset	$PC_{_next} = $ program start
Halt	$PC_{\_next} = PC_{\_current}$
Normal working (default)	$PC_{\_next} = PC_{\_current} + 1$
Jump or call condition satisfied	$PC_{_next} = $ Target address
Interrupt handling	$PC_{_next} = $ interrupt vector entry
Return from call or interrupt	$PC_{_next} = $ Stack top
Watch dog time out	$PC_{_next} = $ program start

certain instruction or pipeline organization. For example, the double-load operation loads two data from memory to RF in one cycle. And the pipeline stage in certain processor is not fixed, thus previous instruction with long pipeline stage might write the RF at the same time as later instruction with short pipeline stage does. In such case, the RF organization becomes complex in that specific control hardware should be developed to handle the data hazards.

General register files are used as the first level computing buffer in processor, while special registers are introduced for specific purposes, such as addressing registers, stack registers, and registers for peripheral. These special registers cannot be located inside the general RF, because they don't need to supply operand data directly for ALU and MAC. In this case, the special registers data processing can be done in such way that: moving the data from special register to RF firstly, then processing the data in RF, and storing the results back to the special registers at last. Thus three clock cycles are needed for one data processing in special register.

Program Counter

The kernel hardware of control path includes PC (Program Counter) and its FSM (Finite State Machine). It updates the PC to address the program memory for next instruction. The state transfer table of the FSM is depicted in Table 2. The initial state of the PC FSM is from system reset, which has the highest priority. When the Call and Return condition is satisfied, the next PC address is firstly reserved in stack, and then jump to the target address. The stack top address is resumed as the next PC address when the subroutine is finished. Today's DSP usually develop zero overlapped hardware loop to support a single-instruction iterative execution. In such a state, the PC FSM holds the current address until the iterative subroutine done. Figure 18 describes a simplified program counter module circuits.

Instruction Decoder

The instruction decoder decodes the machine codes of instruction into several groups of control signals for the pipelined processors. These control signals ensure that all the elements in pipeline and the pipeline itself work as planed. The control signals typically consist of datapath control group (rf_ctrl, alu_ctrl and mac_ctrl),

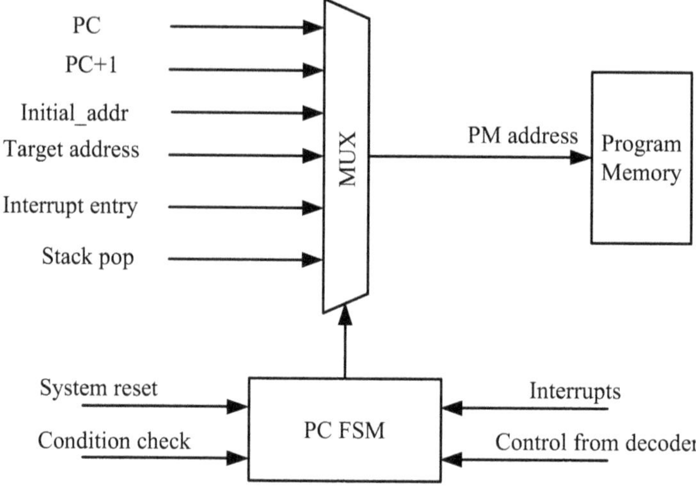

Fig. 18 Program counter module

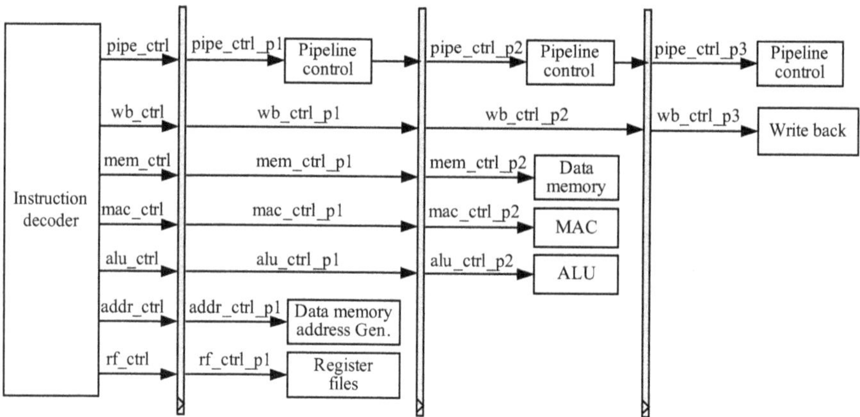

Fig. 19 A simplified instruction decoder circuits

and memory control group (addr_ctrl and mem_ctrl), and pipeline control group (pipe_ctrl and wb_ctrl), illustrated in Fig. 19. Because elements may locate at different pipeline stages, some of the decoded control signals need to be registered for further using. For example, the WB control signals will keep at least 3 stages until the RF write operation really happens.

Pipeline control signals are generated with the purpose to schedule one instruction into different pipeline stage and solve the pipeline hazards. The data forwarding technology can be introduced to effectively handle the pipeline hazards, such as the case of read-after-write, write-after-write and write-after-read. As the pipeline

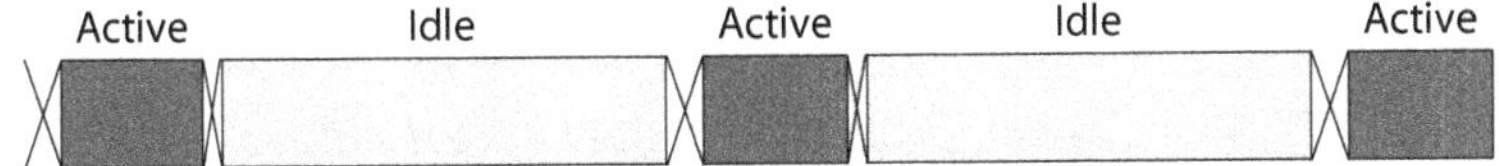

Fig. 20 Working flow of audio-processing algorithms

getting longer, the pipeline hazards get even worse and the control scheme become much more complex, as a result the speedup of the pipeline concept may be eliminated by different pipeline hazards.

This section has introduced the low-power ASIP design flow roughly. The design of ASIP is a hardware/software co-design and iterative process. Architecture customization can be realized by adding customized instructions and hardware accelerators to match the applications. The low-power hardware implementation together with optimized software implementation efficiently using the hardware resource makes the design power optimization. The design of ASIP is also a process of trade-offs between area, performance and power.

2.3 Physical Level Low-Power Techniques

In most of the applications, such as cell phone, audio-processing, video-processing, systems operate in burst mode, which is working in high performance mode for a short time following with a long interval in idle mode. As shown in the Fig. 20, audio-processing algorithms works like this. The working mode is called active state of system, while the idle mode is called standby.

Furthermore, even the system is in active state, it doesn't mean that the system must work in the highest performance all the time. During design stage, the designers must make sure that the DSP can afford the maximum workload. However, the DSP usage is very low during most of the time. So, chance is coming. When workload is not the highest, there is some space for power optimization by doing some trade-off dynamically between performance and power.

According to the description above, the standby mode has gained a lots attention for its large overall power budget. Ideally, the dynamic power consumption during standby period should be reduced to zero through careful design. Furthermore, the static power should also be very small. However, the static power consumption is always hard to cut off with advanced technology scaling. The focus of this section is to discuss several methods to eliminate both dynamic and static power dissipation during standby period.

Clock Gating

All the switches of a subsystem in standby mode are the main source of dynamic power consumption, which is function meaningless, including clock tree and data

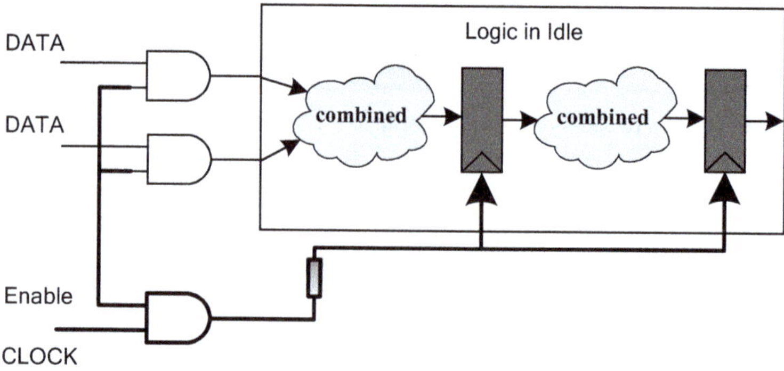

Fig. 21 Structure of the clock gating

propagation. So clock gating technology is implemented with the data isolation, as shown in Fig. 21.

The Structure of Clock Gating includes:

- Isolate the global clock tree from clock tree in idle subsystem.
  With clock tree isolated, no switching activities happen and the clock tree load capacity is eliminated efficiently. Thus, the power consumed by the clock tree can be removed.
- Isolate the input of standby subsystem from the output of the pre-subsystem.
  When the subsystem's clock tree is isolated, the activity of the inputs also brings on some additional power consumption in the combined logic. Isolation between pre-module output and input of this module can efficiently eliminate this part power.

That how and where to implement the clock gating has a vital impact on the sub-module in standby mode.

- Shut down the clock tree of sub-module.
- Shut down the clock tree of the system.
- Shut down the clock tree of the system and the clock generator.

So, kinds of work mode exist because of different management methods for system clock, such as multi-mode supported in today's DSP Core:

- All the sub-modules of the system work
- Part of the clock networks are shut down
- The clock networks of all sub-modules but timer are shut down
- Shut down all the clock network of system

The MSP430 of TI is one of the lowest power consumption MCU. It takes some measures to deal with the wake-up time from standby mode. Its wake-up time is less than 1us, which leaves longer time for sleep mode. It also reduces the power consumptions during the wake-up period. Figure 22 gives its power consumption in standby mode and active mode.

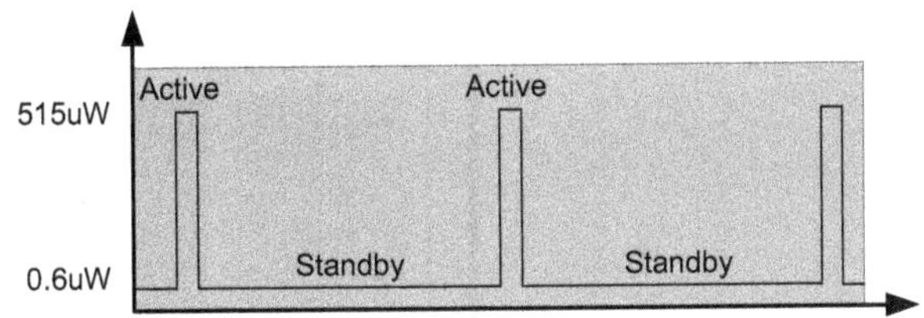

Fig. 22 Power of MSP430 in standby mode and active mode

Clock gating is one of the most widely used low power methods. It is compatible with the standard design flow and EDA tools. In addition, the area penalty of clock gating is negligible. Clock gating technology can be implemented either in RTL level, written by designers, or in gate level, generated by design synthesis tools. Today's logic synthesizer can easily insert clock gating module into design wherever it finds a multiplexer logical structure. However, the tools can't implement global clock gating scheme. Designers can directly instantiate clock-gating logic cell into RTL code to implement global clock gating.

However, it has some impact on the system clock tree, increasing clock skew. Further more, the operand isolation may worsen the delay of the critical path. In addition, clock gating causes the load on the clock network varying dynamically, which may introduce another noise source into system.

The challenge of the clock gating is how and where to implement in the system. According to the usage granularity, there are two methods: grain granularity and coarse granularity. Coarse granularity is the most widely used now. Additionally, whether a sub-module is suitable for using clock gating, it depends on how long the standby mode lasts.

Power Gating

When clock gating is used for the sub-modules in standby mode, the switch power can be reduced nearly to zero. However, the static power remains. The power gating is cutting off the supply of the idle sub-module, reducing the static power to zero. It inserts switch transistor between the global power network and the virtual power network, which is the local power of the idle module. So, when the switch transistor is open, the sub-module works as usual. When the switch transistor is off, the sub-module also is cut off. The structure of the power gating technology includes: the switch network, input and output isolation cell and state retention registers.

As shown in Fig. 23, a normal structure of SOC include: DSP Core, Cache, BUS, peripheral devices, external memory and clock generator. An additional module in the Fig. 23 is the power gating controller, which is especially designed for the power gating. The power network of the system is divided to three parts: VDDSOC, VDDCPU, VDDRAM. VDDSOC is always on, powering the parts which are always working, such as clock generator, power controller and peripheral devices. VDDCPU powers the DSP Core and the Cache. VDDRAM supplies power to the external memory.

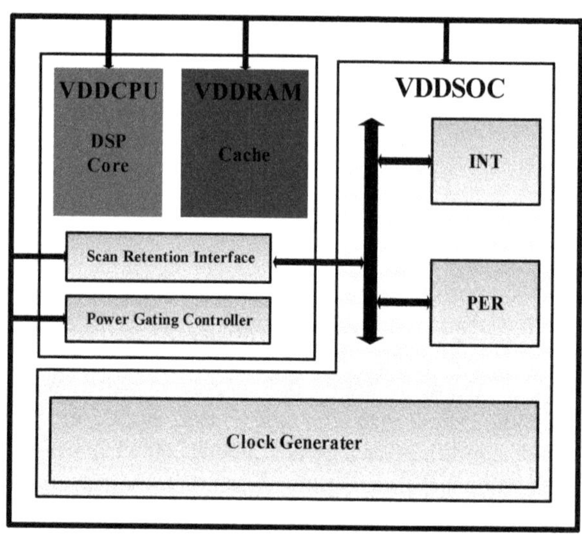

Fig. 23 The Structure of the Power Gating

The working state of a system, which has taken the power gating technology, can be divided to multi-level:

- Shut down the clock of DSP core, however keeping the supply on.
- Cut off the power supply of the DSP core, keeping the key states by the state retention register. Cache works as usual. The system can be awakened in the least time.
- Cut off the power supply of the DSP core, transferring the key states to the external memory through the scan register chains. Cache works as usual. The system can be awakened in the less time.
- Cut off the supply of the DSP core, Cache, transferring the key state to the external memory through the scan chains. When the system awakened, the content of the cache must be rewritten by instructions. The awakening time is the longest.

Power gating is a key method which must be implemented in the ultra low power system. It is compatible with the standard design flow and EDA tools. However, it should be noted that power gating is not suitable for all application. For the application without a significant idle time, the power gating technology may not make any sense. The case may get worse for the power penalty by frequently turning off/on switch transistor. Designer should also notice that the implementation of the power gating makes the synthesis and the backend of the system more complex, especially for the power network and the clock verification.

Comparing to the clock gating, the penalty of the power gating is more serious. Considering the area, the power gating needs many state retention registers and the isolation cells. Sometimes the area of the system using grain granularity can be $2\sim4$ times as before. As for the performance, the introduction of the switch transistors reduces the supply voltage, worsens the critical path.

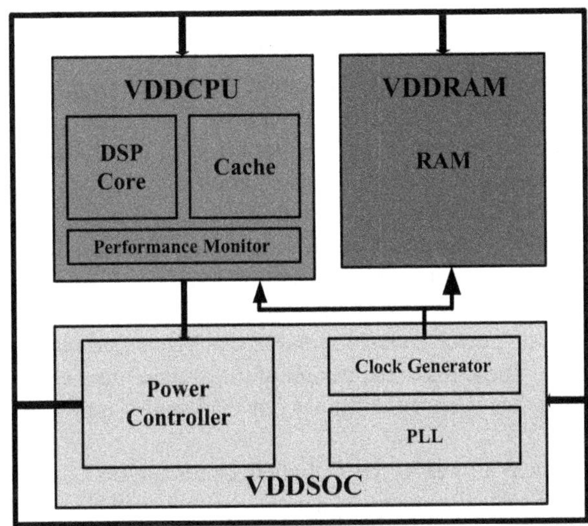

Fig. 24 The structure of the dynamically voltage and frequency scaling

There still exist some challenges for the power gating technology. The awaking time and the voltage pulse are both very important to the switch network. However, the improvement of one always worsens the other. In addition, the power gating makes it more difficult to the verification of the system. Lastly, it is very difficult to estimate the power consumed during the awaking time. So, whether power gating is suitable for a sub-module is a result of estimation of the idle time and the power consumption during the awaking time.

Dynamic Voltage and Frequency Scaling

DVFS is a method that monitoring the workload of the system, then dynamically managing the voltage and the frequency scaling, regulating the system state, to achieve the aim of power reduction. DVFS needs to divide the system into several power domains. Each works with an independent power supply. A performance monitor module is added, monitoring the workload of the system, estimating the real delay needed of the critical path, then transmitting the messages to the voltage management module and the clock generate module, regulating the voltage supply and the clock frequency of the sub-module.

The structure of the DVFS is shown in the Fig. 24. The system is divided into several power domains: (1) DSP Core, Cache, CPU performance monitor module, powered by VDDCPU; (2) Memory, powered by VDDRAM; (3) BUS, BUS performance monitor module, clock generate module, self-adjust voltage controller module, PLL, powered by VDDSOC.

Supposed the system works in a normal voltage and clock frequency. When the performance monitor module detects the fact that performance of the system can not meet the need of the workload, the voltage of the CPU should increase. Firstly, lock the frequency of the CPU to the current frequency. Then, increase the voltage of the CPU until the performance need of the workload can be met. Lastly, transmit the frequency to the target frequency. Vice, versa.

DVFS technology is compatible with the design flow and EDA tools, widely used in the IC design. It makes good use of the performance needs of the system workload, achieving the trade-off between performance and power in real time, resulting in the power reduction. More important, the implementation of the DVFS has little impact to the highest performance of the system.

However, When implementing the DVFS in a system, performance monitor module, power manage module, clock generate module should be designed especially. These make the design of the system more complex. Additionally, the level-shifter between the different power domains may worsen the delay of the critical path.

There also exist some challenges for the DVFS. How to estimate the workload of the system accuracy has a greatly impact on the efficiency of the DVFS. How to generate suitable voltage and frequency to the system is another serious problem, which also has a greatly impact on the optimum system state.

3 Low Power ASIP for Audio Application

An ultra low power ASIP called FlexEngine is specially developed for audio application, such as audio encoding and decoding, sound enhancement algorithms, noise reduction. The FlexEngine is a 24-bit low power DSP core, which has optimized instruction set and high-efficient microarchitecture. As FlexEngine DSP is a well balanced core, achieving low power, high efficiency, and low gate count, many audio domain implementations can benefit from it.

3.1 Overview

The microarchitecture of FlexEngine DSP is described in Fig. 25. The processor core introduces distributed Harvard memory architecture, using one program memory and two separate data memories. The pipeline scheme partitions each instruction into 5 sub operations, achieving parallel execution for each cycle. The critical path for each pipeline stage is well balanced to gain a maximum clock frequency. Customized instructions are added to accelerate the audio processing. Meanwhile, the hardware of the accelerator is carefully designed with low power and low gate count. Zero overlapped hardware loops is developed to eliminate the subroutine jump overhead. The AGUs (Address Generation Units) offer extensive addressing mode for regular embedded signal process. The advanced data forward scheme im-

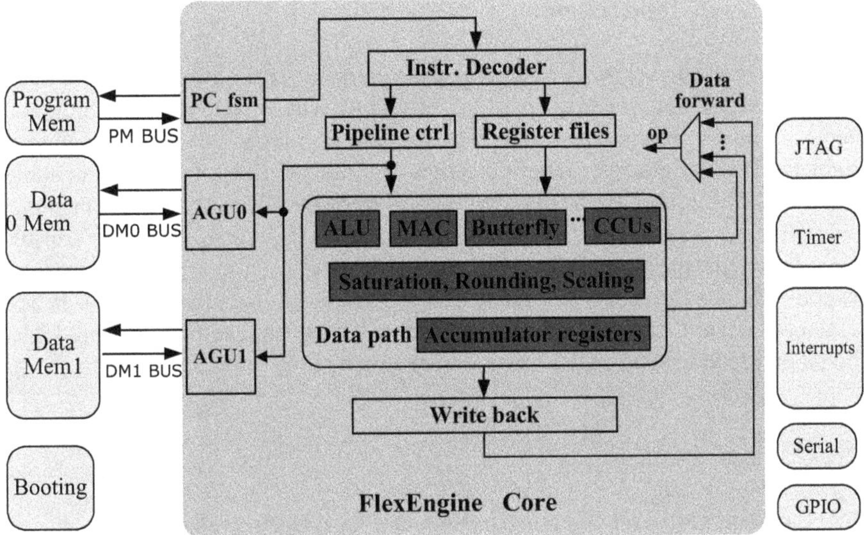

Fig. 25 FlexEngine DSP architecture

proves the execution parallelization by hiding various data hazards. The data path consists of ALU, MAC and Customized Computation units.

Hardware features include:

- Distributed Harvard memory architecture
- Full 24-bit data paths, 64K words address space for program and data memories
- RISC instruction set, suitable for both control and computation
- Zero overhead loops
- Various addressing modes with modulo protection, bit reversal
- Data forward scheme
- Four 56-bit accumulators
- Supporting Saturation, Rounding and Scaling operations
- Customized Computation Units for audio acceleration
- Peripheries: Timer, Interrupts, Serial ports, GPIO
- JTAG: breakpoint, single step, run/stop, memory read/write

3.2 Low Power Optimization

The design of FlexEngine DSP follows the approach described in Sect. 2.2. In order to further improve the energy-efficiency, some special low power optimization methods are developed.

Memory Power Optimization

According to the power analysis reports, the memories consume about 50 % of total system power. A new structure called loop buffering is developed to reduce program memory's power dissipation.The loop buffer typically has a 128-word SRAM and stores the DSP program for iterative computing. For all iterative loops computing tasks, the FlexEngine core fetches instructions from the loop buffer rather than the large program memories. Thus, a great part of program memory's power can be saved. To reduce the power of data memories, dividing the large memory space into several little storage banks can make sense. Then, only one memory bank is accessed whenever there is a read/write operation. Additional address decoding logic should be added to support the new memory structure.

Clock Gating

Both local and global clock gating schemes are introduced to eliminate the meaningless switch. The global clock gating is implemented manually during RTL coding. The sleep instruction will transparently shut down the clock switch of the entire core, leaving only periphery active. Moreover, the clock of each computing unit (such as ALU, MAC, and CCUs) is on only when the current instruction is truly implemented on these units. The local clock gating is probably implemented by today's synthesizer. Designers need to turn on the clock_gating option in the synthesis scripts. It should be noted that with a perfect clock gating scheme, the power of DSP is mainly determined by the current workload rather than the clock frequency.

Low Voltage

Reducing the supply voltage will slow down the DSP performance, since a large number of gates suffering from a long delay. In some cases, the system need power budget below 1 mW, even μW in certain case. Then, low supply voltage is the significant to realize that goal. Besides three regular library corners offered by IC foundry, low voltage library corner is modeled by EDA tools. Both the logic and physical designs are based on the new low voltage library. The FlexEngine core is implemented both in normal voltage and low voltage in CMOS 130 nm technology, with power analysis results 70 μW/MHz@1.2 V and 34 μW/MHz@0.8 V respectively.

The FlexEngine DSP is a well balanced core for ultra low power applications. The targeted applications include hearing aid devices, portable audio players, cell phones, and headsets. The FlexEngine core can also be an IP or co-processor executing intensive DSP tasks in SoC, such as speech enhancement, noise reduction, and audio encoding/decoding.

References

1. Michael Keating, David Flynn, Robert Aitken, *et al.* Low Power Methodology Manual: For SoC Design[M]. New York: Springer, 2007.
2. Jan Rabacy. Low Power Design Essentials[M]. New York: Springer, 2009.
3. Jan M. Rabacy, Anantha Chandrakasan, Borivoje Nikolic. Digital Integrated Circuits: A Design Perspective (Second Edition) [M]. Beijing: Tsinghua, 2004.
4. Hao Dongyan, Zhang Ming, Zheng Wei. Design Methodology of CMOS Low Power. IEEE, 2005.
5. Bo Zhang, Liping Liang, Xingjun Wang. A New Level Shifter with Low Power in Multi-Voltage System. IEEE, 2006.
6. Stephanie Ann Augsburger. Using Dual-Supply, Dual-Threshold and Transistor Sizing to Reduce Power in Digital Integrated Circuits[M. S. thesis]. Berkeley, The USA: University of California at Berkeley. 2002.
7. Danny P. Riemens. Exploring suitable adder designs for biomedical implants[M. S. thesis]. Delft, The Netherlands: Delft University of Technology. 2010.
8. Chris H. Kim and Kaushik Roy. Dynamic V_{TH} scaling scheme for active leakage power reduction [C]. Proceedings of the 2002 Design, Automation and Test in Europe Conference and Exhibition (DATE.02), 2002.
9. David Duarte, Yuh-Fang Tsai, Narayanan Vijaykrishnan, *et al.* Evaluating Run-Time Techniques for Leakage Power Reduction[C]. Proceedings of the 15th International Conference on VLSI Design (VLSID.02), Bangalore, India, 2002.
10. A. Correale, Overview of the Power Minimization Techniques Employed in the IBM PowerPC 4xx Embedded Controllers, ISLPED, 1995, pp. 75–80.
11. Jun Chao, Yixin Zhao, Zhijun Wang, *et al.* Low-power implementations of DSP through operand isolation and clock gating. IEEE, 2007.
12. N. Banerjee, A. Raychowdhury, K. Rey, *et al.* Novel Low-overhead Operand Isolation Techniques for Low-Power Datapath Synthesis. IEEE, 2006.
13. Qing Wu, Massoud Pedram, and Xunwei Wu. Clock-gating and its application to low power design of sequential circuits. IEEE, 2000.
14. Michael De Nil, Lennart Yseboodt, Frank Bouwens. Ultra Low Power ASIP Design for Wireless Sensor Nodes. IEEE, 2007.
15. Kaijian Shim, David Howard. Challenges in sleep transistor design and implementation in low-power designs. DAC 2006, San Francisco, California, USA, July 24–28, 2006.
16. Arne Martin Holberg, Asmund Saetre. Innovative Techniques for Extremely Low Power Consumption with 8-bit Microcontrollers. Atmel Corporation White Paper, 2006.
17. Nariman Moezzi Madani, Naser Masoumi. A new optimization method for CTMDP system-level power management techniques. IEEE, 2004.
18. Manoj Kumar Jain, M. Balakrishnan, Anshul Kumar. ASIP Design Methodologies: Survey and Issues. In Int. Conf. on VLSI Design, Jan. 2001.
19. Tilman Glökler, Heinrich Meyr. Design of Energy-efficient Application Specific Instruction Set Processors, 2004, Kluwer Academic Publishers.
20. M. Hohenauer, H. Scharwaechter, K. Karuri, O. Wahlen. A methodology and tool suite for C compiler generation from ADL processor models. In Proceedings of the Conference on Design, Automation & Test in Europe (DATE), 2004.
21. Tensilica Inc.: http://www.tensilica.com.
22. CoWare Inc.: LISATek product family, http://www.coware.com.
23. Dake Liu. Embedded DSP Processor Design, 2008.
24. J. Henkel, S. Parameswaran. Designing Embedded Processors: A Low Power Perspective, 2007, Springer Publisher.
25. VIVEK TIWARI, SHARAD MALIK, ANDREW WOLFE. Instruction Level Power Analysis and Optimization of Software. Journal of VLSI Signal Processing.1996.
26. Paul Eric Landman. Low-Power Architectural Design Methodologies. 1994. pp. 63–68.

Ultra-Low Power Transceiver Design

Hanjun Jiang, Nanjian Wu, Baoyong Chi, Fule Li,
Lingwei Zhang and Zhihua Wang

1 Introduction to Ultra-Low Power Transceiver for Medical Applications

Many emerging medical application systems require the wireless communication functions. Nowadays, wireless medical application for sensing purposes such as the long term electrophysiology signal (ECG, EEG, etc.) monitoring [1], the digestive tract imaging with the wireless endoscope capsule [2], the fetal heart sounds tele-examination [3] and the real-time pressure sensing in artificial joint replacement surgeries [4] have gained many remarkable achievements. Also, the wireless medical applications for intervention purposes such as neural stimulation for cochlear implants [5] and insulin delivery with closed-loop control [6] have preliminarily demonstrated the clinical practice value.

A wireless medical application system is usually composed of the local side and the remote side, as shown in Fig. 1. The local side includes one or multiple sensing or intervention devices (SID) as the nodes of the wireless body area network (WBAN), and a WBAN hub which connects to the SID nodes wirelessly to collect sensing data or to transfer commands. The WBAN hub also establishes links with the wide area networks (WAN) or the telecommunication system in the remote side to transfer the health information to doctors for diagnosis or automatically call for rescue if necessary. The wireless communication between the WBAN hub and nodes is the so-called short-range communication, while the wireless communication between the local side and the remote side is long-range. The wireless communication discussed in this chapter will be limited to the short-range communications.

H. Jiang (✉) · B. Chi · F. Li · Z. Wang
Institute of Microelectronics, Tsinghua University, Beijing, China
e-mail: jianghanjun@tsinghua.edu.cn

N. Wu
Institute of Semiconductors, Chinese Academy of Sciences, Beijing, China

L. Zhang
Galaxycore Shanghai Ltd. Corp., Shanghai, China

N. N. Tan et al. (eds.), *Ultra-Low Power Integrated Circuit Design*,
Analog Circuits and Signal Processing 85, DOI 10.1007/978-1-4419-9973-3_6,
© Springer Science+Business Media New York 2014

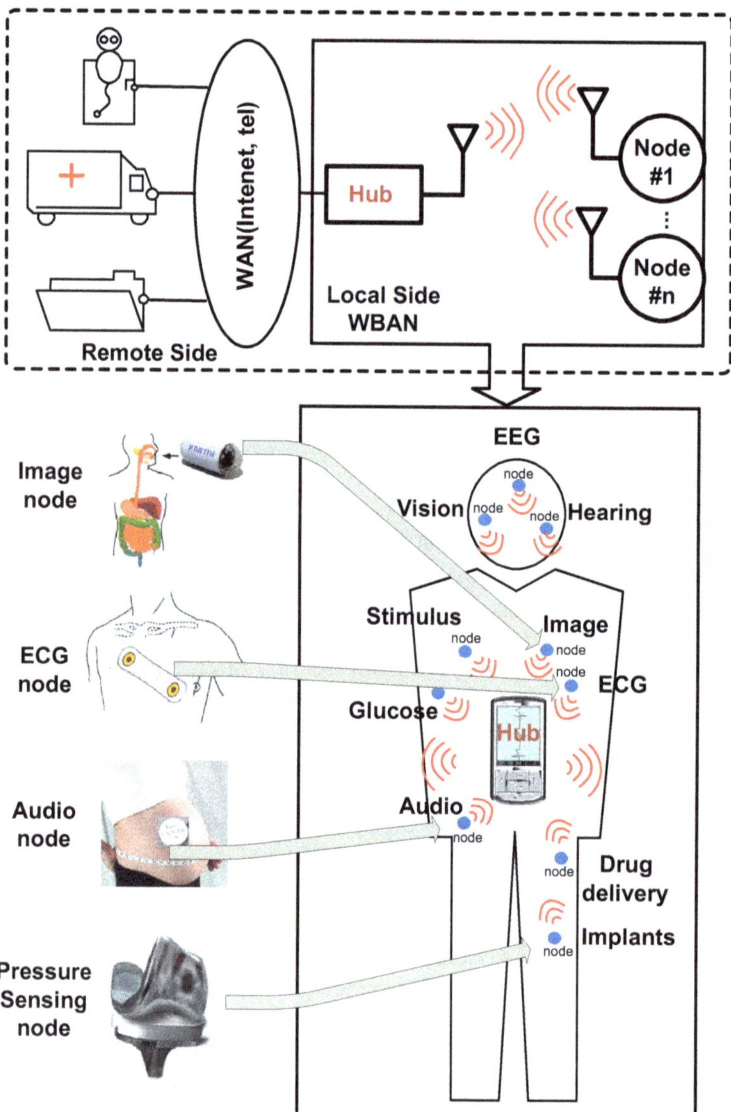

Fig. 1 Typical wireless medical applications that requires short-range wireless transceivers

As shown in Fig. 2, a WBAN hub is usually composed of a MCU for system control, a power management unit (PMU), a memory section for program/data storage, a short-range transceiver to communicate with the SID's, and the interface to Ethernet, wifi, GPRS, etc., and the series port such as the USB port to PC and/or other devices. A SID node, as shown in the right part of Fig. 2, is usually composed of light-weight MCU, an integrated PMU, and most importantly, the interfaces to

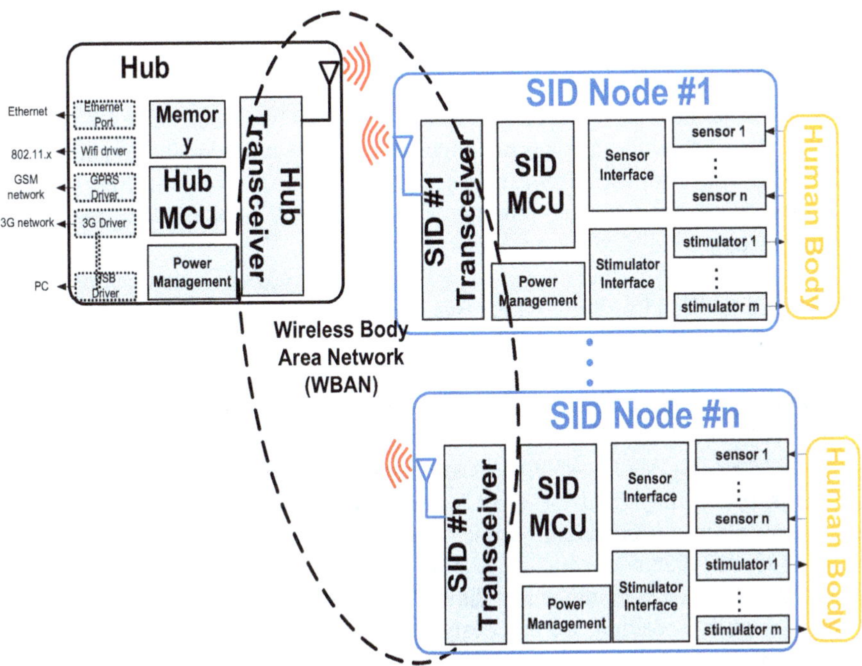

Fig. 2 Function blocks of a typical WBAN system with SID nodes and a hub

biomedical sensors and bio-stimulators. The transceivers between the hub and SID node's carry out the wireless body area communication. In many situations, the hub works with a set of SID nodes for one personnel and a half-duplex single-hop star topology network can be adopted [7].

Both for the SID nodes and the hub in the emerging wireless medical applications, one of the most important function blocks is the short-range WBAN transceiver. According to the IEEE 802.15.6 WBAN standard released in February, 2012, the short-range low-power WBAN is operated on, in or around the human body (but not limited to humans) [8]. There are three types of wireless transceivers that can be used for WBAN, namely, the narrow band (NB), the ultra wideband (UWB) and the human body communications (HBC) transceivers. Though the latter two types still require some time to validate some critical issues such as safety, form factor when considering the external components, etc., the NB transceivers have proven its effectiveness in many real applications. The remaining part of this chapter will focus on the NB transceivers for WBAN.

From the viewpoint of implementation, the most critical constraint for SID nodes is the form factor. Since the SID nodes are usually powered by tiny coin batteries, the ultimate essential to for SID node circuit design is ultra-low power consumption (ULP). Every part of a SID node has to be ULP, including the sensor interface, stimulator interface, flow controller and the transceiver. All the circuit parts should be optimized in terms of power consumption, according to the specific application.

The ultra low power design is especially important to design the SID node transceiver, since the wireless transceiver may contribute the highest power consumption p to SID nodes.

On the other hand, a WBAN hub is usually powered by rechargeable Li-ion batteries, and the power limitation is not quite critical. However, in many situations the hub should work with various SID nodes requiring different communication speeds, link budgets, etc., so the most important challenge for WBAN hub transceiver design is its compatibility to various types of SID node.

To build a WBAN, there are quite a few existing protocols that can be used, such as IEEE 802.11.6, 802.15.4/ZigBee and 802.15.1/Bluetooth. In the recently released IEEE 802.11.6 standard, the physical (PHY) layer defines a narrow band (NB) mode compliant device (hub/node) shall be able to support at least one of the following frequency bands: 402–405, 420–450, 863–870, 902–928, 950–958, 2,360–2,400 and 2,400–2,483.5 MHz. Before the 802.15.6 standard was released, IEEE 802.15.4/ZigBee low-rate wireless personal area network (LR-WPAN) standard was also widely used for WBAN devices [9]. The IEEE 802.15.4 standard PHY layer defines it operating on one of the three possible unlicensed frequency bands: 868, 902–928 and 2,400–2,483.5 MHz [10].

Another solution is to implement the WBAN transceivers using a dedicated protocol. Generally speaking, we can design a set of dedicated SID node transceivers optimized for different application purposes, and a powerful hub transceiver compatible with those SID node transceivers. The SID node transceiver should be optimized for different applications, such as ECG, pulse rate, EEG, temperature monitoring, and we can choose the most suitable transceiver data rate, transmitter (TX) power level, receiver (RX) sensitivity and etc. for a specific application with the power consumption as the optimization goal. A single SID node transceiver will only choose one frequency band, and has a fixed data rate and simple but effective modulation type. On the other hand, the hub transceiver may supports the body area wireless communication covering the 400, 900 MHz and 2.4 GHz ISM band, and the transceiver speed and modulation type is programmable to cover different data rate speed required for different SID nodes. The power consumption imposed on the PCS SoC is not critical, but we still need to do the best to lower its power consumption. In a word, a multi-band multi-rate multi-mode low-power transceiver will be designed for the WBAN hub, while a set of ultra-low-power transceivers with fixed frequency band, data rate and modulation type will be designed for specific SID applications.

One significant design requirement is that the SID nodes have to be implemented with small form factors, especially for those implantable medical devices (IMD's). Therefore the SID transceiver should be designed with high integration level to make the number of external components as few as possible and the size of external components as small as possible. For example, the designers should try the best to include the antenna matching network inside the chip and provide the automatic antenna tuning on-chip such that the inductors and capacitors for antenna impedance matching can be saved. Another observation is that the designers should make the effort to implement the power management unit on-chip such that the SID node does not need any monolithic DC-DC switch regulators or low-dropout regulators (LDO's).

In the remaining part of this chapter, three CMOS WBAN transceivers with dedicated protocols will be discussed to show some circuit design techniques that can be used for ultra-low-power/low-power wireless transceivers. The first design is a 2.4 GHz transmitter for SID nodes, and the second one is a 400 MHz transceiver optimized for implantable medical devices. The third design is a low-power multi-band multi-mode transceiver that can be used for WBAN hubs with high compatibility.

2 2.4 GHz Transmitter for WBAN Nodes

In this section, the design example of a 2.4-GHz energy efficient transmitter for WBAN SID nodes will be discussed. A low-power, low hardware complexity ON-OFF keying/frequency-shift keying (OOK/FSK) modulation scheme is adopted in the design. They are suitable for the design of a smart, compact, and low-power medical sensor nodes which makes the medical monitoring available anytime and anywhere. The transmitter combines the VCO direct modulation mode and the PLL-based mode to make good use of their advantages. In particular, the data rate of the PLL-based mode can be increased by a frequency presetting technique without increasing the loop bandwidth [11]. The frequency presetting can directly preset the frequency of VCO with small initial frequency error and avoid the tradeoff between the lock-in time and the phase noise or spurs. The technique can reduce the lock-in time and increase the frequency switching speed greatly so that the data rate of the transmitter can be increased with low power consumption.

A digital processor is designed to control the operation mode of the transmitter and to preset the PLL or VCO output frequency. Furthermore, ultra-low- power nonvolatile memory (NVM) based on a standard CMOS process is integrated in the transmitter to store the presetting signals and calibration data obtained in the chip testing process. This avoids the repetitive calibration process and saves the power consumption in practical applications. The class-B PA is also adopted to save the energy for transmitting. As a result, the transmitter can achieve OOK/FSK modulation with a high data rate in a high energy efficiency way. The overall architecture, the circuit design details as well as the measurement results will be presented in the following subsections.

2.1 *Transmitter Architecture*

Figure 3 shows the transmitter block diagram. It consists of four major function blocks: a PLL frequency synthesizer with the frequency presetting function, a class-B power amplifier (PA), a digital processor, and NVM. The transmitter can operate

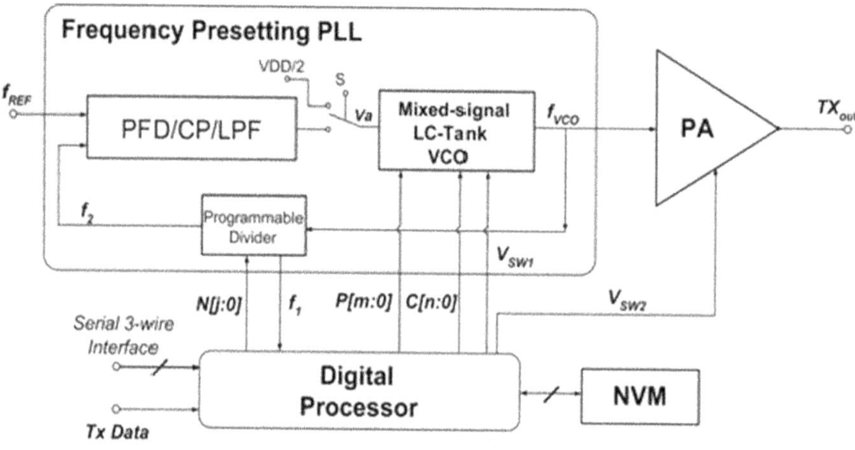

Fig. 3 2.4 GHz transmitter architecture

in two modes: (1) PLL-based mode and (2) VCO direct modulation mode with OOK/FSK modulation.

The PLL frequency synthesizer is composed of five function blocks: the phase-frequency detector (PFD), the charge pump (CP), the second-order low-pass filter (LPF), the mixed signal LC-tank VCO and the programmable divider.

The VCO oscillation frequency is determined accurately by the output signals $C[n:0]$ and $P[m:0]$ of the digital processor and is tuned finely by the output signal of a low-pass filter (LPF) through the presetting module. The frequency presetting technique can accurately preset the frequency of the VCO with a small initial frequency error and reduce the lock-in time directly so that the frequency switching speed and the data rate of the transmitter can be increased greatly with low power consumption. The PFD uses a three-state phase/frequency detection scheme. Four inverters are inserted in the reset path as the delay cell to avoid the dead-zone problem. The CP is based on the low-power source switch topology. The LPF is a second order on-chip RC filter. The divider is a conventional one that consists of a dual-modulus pre-scaler (DMP) and a pulse-swallow counter. It can generate two frequency signals and for frequency calibration and PFD, respectively.

The digital processor can perform the following operations. First, the processor measures the output frequency of DMP and calibrates the relation between the VCO output frequency and the digital presetting signals $C[n:0]$ and $P[m:0]$ automatically when set to VDD/2. Then, it stores the calibration data into the NVM. In practical applications, the processor can accurately preset the output frequency of the VCO by the presetting signals $C[n:0]$ and $P[m:0]$ so that the frequency of the PLL synthesizer can be rapidly switched. Furthermore, the processor receives a baseband signal from the baseband processor and generates the signal V_{SW1} and V_{SW2} for the VCO and PA to perform OOK/FSK modulation.

The PA is a push-pull class B buffer amplifier with little dc power consumption. It consists of a class B amplifier and a buffer amplifier. The class B amplifier looks like a CMOS inverter, but the transistors in PA are biased around the threshold voltage to achieve high efficiency. The buffer amplifier is self-biased to reduce the second-order harmonic. Some complementary switches are adopted to reduce the rising time and falling time of the PA.

The ultra-low-power NVM integrated in the transmitter is based on FN tunneling phenomenon with extremely low current density. It stores the control signals and calibration parameters obtained in the chip testing process. This avoids the repetitive calibration process and saves the power consumption in practical applications.

The operation flow of the transmitter is described as follows. When the transmitter starts up or receives a reset signal for the first time, it first performs frequency auto-calibration. The digital processor measures the output frequency of VCO and calibrates the relation between the output frequency and the presetting digital signals C and P automatically. When the calibration process is finished, the presetting signals and calibration data obtained will be stored into the NVM. Next, the transmitter transmits the sampled information to the external WBAN hub. The digital processor reads the required presetting signals C and P from the NVM directly to preset the frequency of the VCO. In the VCO direct modulation mode, the PLL loop is cut off and is biased to VDD/2. In the PLL-based mode, after the output frequency of VCO is preset with a very small initial frequency error, the output voltage of the LPF precisely tunes the frequency of the VCO. Therefore, the synthesizer can settle down in a very short time. Its lock-in time almost does not depend on frequency step, process variation, device parasitic effect, and temperature. OOK modulation can be achieved by switching the control signal and directly by the baseband data while FSK modulation can be realized by changing two different groups of presetting signals $C[n:0]$, $P[m:0]$ of VCO and divide ratio $N[j:0]$ of DMP.

Frequency Presetting Method

The precision of the VCO output frequency and the lock-in time of PLL are important design parameters in the VCO direct modulation mode and the PLL-based mode, respectively. The frequency presetting technique proposed in [10] can be used to overcome the difficulties. The mechanism where the frequency presetting method reduces the lock-in time of the synthesizer significantly is as follows. The lock-in time of the frequency synthesizer is defined as (1):

$$T_l = \frac{-\ln\left(\dfrac{tol}{\Delta f} \times \sqrt{1 - \zeta^2}\right)}{\zeta \times \omega_n} \tag{1}$$

The factor tol is the acceptable frequency error, Δf is the initial frequency error, ζ is the damping factor, and ω_n is the natural frequency of the PLL loop. Since the lock-in time depends on the magnitude of the initial frequency error, the lock-in time can be shortened as long as the initial frequency error can be reduced. In order to reduce the frequency error, a novel mixed-signal LC-tank VCO is designed. The oscillation frequency of the mixed-signal VCO can be preset accurately by the digital signal and finely tuned by the output signal of the LPF in the PLL loop. The frequency presetting technique is used to preset the frequency of the VCO to the target frequency with a small initial frequency error by presetting the signals of the digital processor.

Nonvolatile Memory

For the medical applications, the temperature of the human body is almost constant. And the supply voltage is unchanged with the on-chip voltage regulators. So the frequency calibration just needs to be done only once after the SID node is worn or implanted.

The NVM can store the presetting signals and calibration parameters when powered down so that the repetitive calibration process is avoided and the power consumption is saved in practical applications. The typical EEPROM or FLASH memory requires additional masks so that even a small size of embedded memory brings additional cost for the entire system.

Furthermore, these kinds of memories usually consume a lot of power. Recently, an ultra-low-power NVM with a high efficiency charge pump circuit has been designed [12]. The memory can be integrated in a standard CMOS process and it is based on FN tunneling phenomenon with extremely small current density. Therefore, it is suitable for the targeted ultra-low-power medical applications.

2.2 Circuit Design Details

Low-Power Mixed-Signal LC-Tank VCO

Figure 4 shows the proposed mixed-signal VCO. It consists of an LC-tank VCO and a presetting module, as shown in Fig. 4a, b, respectively. The LC-tank VCO adopts complementary-type PMOS and NMOS to reduce the current needed for oscillation. In order to reduce the power consumption, we select a large inductor L with a high Q value to increase the loop gain of VCO. The digital processor generates two terms of digital signals C and P to control the output frequency of VCO. The digital signal $P[3:0]$ is adopted to control capacitance of the LC tank and generate 16 overlapped discrete tuning curves to increase the desired frequency tuning range and lower VCO gain K_v [13]. Smaller will benefit phase noise performance and it also improves the resolution of frequency presetting. The digital signal $C[5:0]$ controls the presetting module with the output signal V_a of LPF to produce a VCO control signal V_c. Thus, the signals C and P can accurately preset the output frequency of the VCO. The signal V_a finely tunes the frequency. In our design, V_a is almost constant

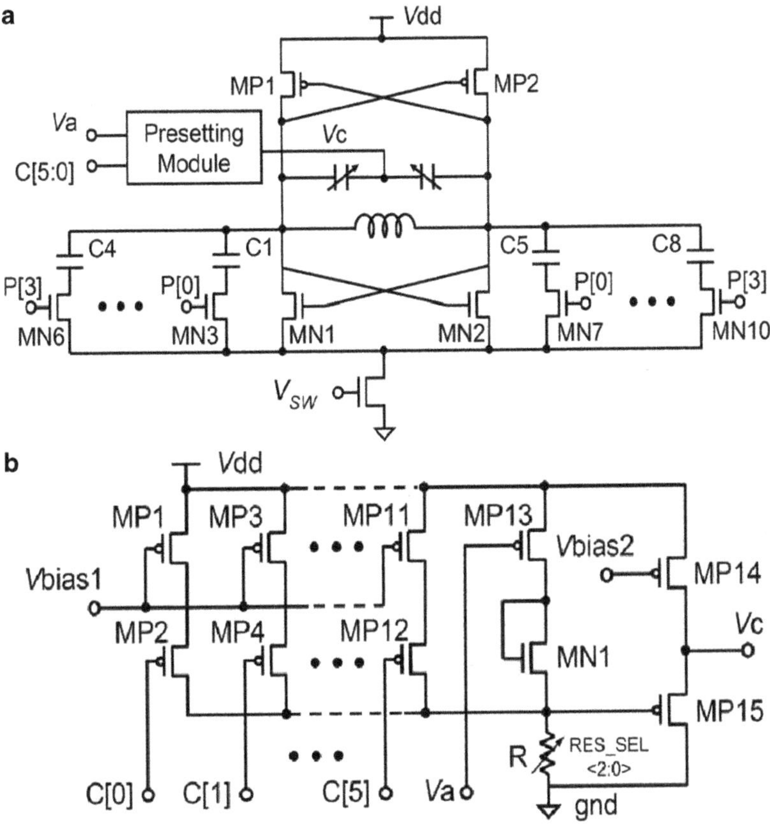

Fig. 4 Mixed-signal VCO with a presetting module. (**a**) Top architecture. (**b**) Presetting module

and equal to VDD/2. Figure 5 shows the measured and simulated dependence of the VCO output frequency on the presetting signals *P* and *C* at 0.9 V.

The presetting module is a mixed-signal circuit. It consists of a series of parallel current sources, switches, resistors, and a source follower. MP1, MP3, and MP11 transistors constitute a series of current sources with a different ratio of their currents: $2n$ (n=1, 2, 3, 4, 5, 6), respectively. The 6-b frequency presetting digital signal *C[5:0]* controls the current sources on/off by the switches MP2, MP4, and MP12, respectively. The linearity between the VCO frequency and the presetting signal *C[5:0]* code can be ensured by carefully matching the design of the current sources, especially the layout design. The Monte Carlo simulations are used to guarantee that the frequency error can always be below the least-significant bit (LSB) of the C code settings across process and temperature. The value of resistor R is tuned by the 3-b digital signal *RES_SEL[2:0]* to make the presetting frequency more accurate. When a digital signal *C[5:0]* is inputted into the presetting module, the module produces the voltage V_c by the source follower, consisting of MP14 and

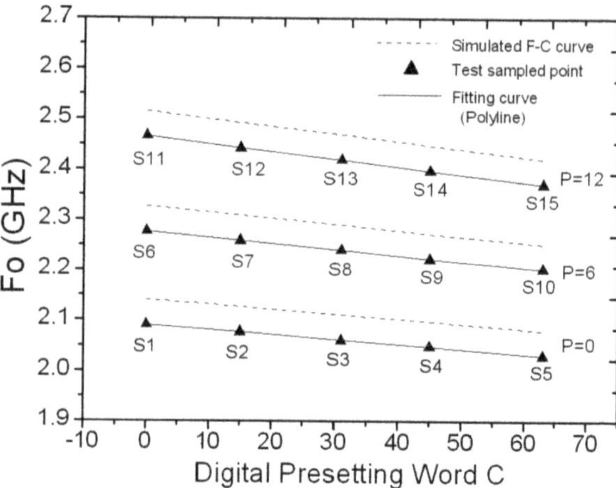

Fig. 5 Output frequency Fo versus presetting signal *C[5:0]*

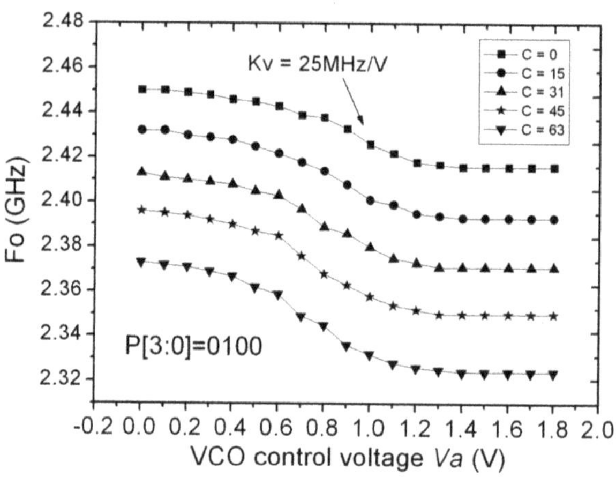

Fig. 6 Transfer characteristics of the proposed VCO

MP15 to preset the frequency of the VCO with a small frequency error. And the output signal V_c of the LPF accurately tunes the frequency of the VCO by adjusting the current through MP13. Figure 6 shows the measured results of the transfer characteristics of the proposed VCO with a different frequency presetting signal C when the signal *P[3:0]* is set to "0100." The measured K_v is about 25 MHz/V near the 0.9 V as shown in Fig. 6.

Frequency Auto Calibration

Figure 5 shows the dependence of the VCO presetting frequency on the presetting signals C and P at V_a=0.9 V. The simulated result shows the good linear relation between the presetting frequency and the signal C under three P signals. But the measured result shows that the dependence of the VCO presetting frequency on signal C deviated from the simulation result due to process variation and the device parasitic effect. The idea of calibrating the switched capacitors to achieve the reduced initial frequency offset is proposed in [14].

In practical applications, the digital processor can automatically calibrate the relation between the presetting frequency and the signals C and P in the calibration mode by the following algorithm. The calibration algorithm is based on a frequency sampling and linear interpolation. First, as shown in Fig. 3, the PLL is opened by the switch S and the input of VCO is biased to 0.9 V. Then, the output frequency f_{VCO} of the VCO is divided by the DMP to generate the signal f_1 around 80 MHz which will be sampled by the digital processor with the reference clock, which is 1 MHz. Thus, the dependence of the VCO presetting frequency on the signal C under different P signals is automatically measured by the frequency sampler. For every P signal, five points are sampled. For example, the sampling points S_i are shown in Fig. 5. Next, the processor produces the measured data to form a fitting curve by the linear interpolation module. After this operation, the value of signals C and P for the target frequency can be calculated from this fitting curve. The fitting curve will be stored into the NVM. Furthermore, the transmitter can be remotely controlled to perform the calibration process again in case the supply voltage and temperature changed.

The error of the frequency calibration mainly results from the error of the frequency sampling and linear interpolation by the digital method. The calibration resolution with 1-MHz reference frequency is discussed as follows. We assume that the target frequency F_T is between the actual frequencies F_{Si} and F_{Si+1} at the sampled points S_i and S_{i+1}. The frequencies obtained by sampling at the points S_i and S_{i+1} are $F_{Si}+\Delta F_i$ and $F_{Si+1}+\Delta F_{i+1}$, respectively, where ΔF_i and ΔF_{i+1} are sampling errors and $|\Delta F_i|, |\Delta F_{i+1}| < 1$ MHz (due to the sampling with a 1-MHz reference clock). The signals C corresponding to points S_i and S_{i+1} are C_{Si} and C_{Si+1}, respectively. Then, the calculated signal C and the calibrated frequency (including error) for target frequency are noted as C_T and $F_T+\Delta f$, respectively. They can be expressed according to linear interpolation in

$$C_T = C_{Si} + \frac{F_T - (F_{Si} - \Delta F_i)}{(F_{Si+1} + \Delta F_{i+1}) - (F_{Si} + \Delta F_i)} \cdot (C_{Si+1} - C_{Si}) \tag{2}$$

$$F_T + \Delta f = \frac{C_T - C_{Si}}{C_{Si+1} - C_{Si}} \cdot (F_{Si+1} - F_{Si}) + F_{Si} \tag{3}$$

Combining (2) and (3), we can obtain

$$F_T + \Delta f = \frac{F_T - (F_{Si} - \Delta F_i)}{(F_{Si+1} + \Delta F_{i+1}) - (F_{Si} + \Delta F_i)} \cdot (F_{Si+1} - F_{Si}) + F_{Si} \tag{4}$$

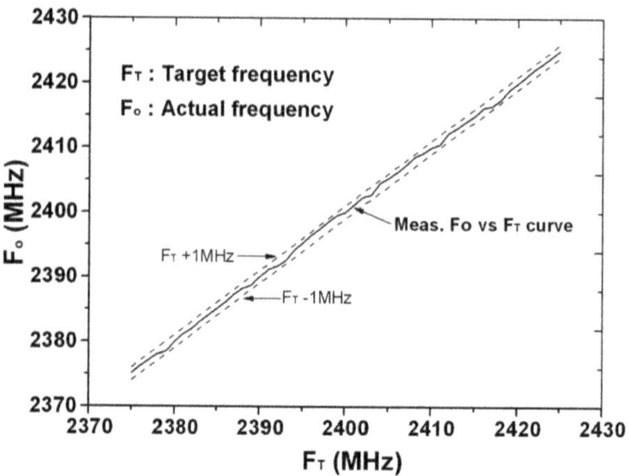

Fig. 7 Measured calibrated frequency

It can be proved that (4) has the maximum value F_{T+1} when $\Delta F_i = -1$ and it has the minimum value F_{T-1} when $\Delta F_i = \Delta F_{i+1} = 1$. Therefore, we obtain $|\Delta f| < 1$ MHz, which means the resolution of the calibration is 1 MHz. Higher frequency resolution could be achieved by longer counting time. And the error can also be significantly reduced if the reference clock is first divided and used for frequency detection during the calibration process. But they all increase the energy for calibration. The measured error between target frequency and actual frequency of VCO of this calibration algorithm is within 0.6 MHz, which is shown in Fig. 7. The measurement results indicated that the probability that (4) has a maximum and minimum value is very small. This is enough for the PLL-based operation mode since the typical lock-in time is several microseconds. And it is also enough for the VCO direct modulation mode because the frequency error can be allowed by the input bandwidth of the commonly used receiver.

Class B Power Amplifier

The class E PA is more widely used to improve the efficiency over other linear-mode PAs. However, the driving stage usually consumes a lot of power to make the output stage switching lossless, which cannot be affordable in the medical applications. And the output matching network is relatively complex which cannot be accepted for miniature SID nodes. Here, the class B PA with a simple matching network and reduced output power is adopted in the design, which is shown in Fig. 8. The PA is a push-pull class B buffer amplifier that mainly consists of transistor M1 and M2. The complementary ac-coupled NMOS/PMOS approach was used in the preamplifier portion of the PA in [15]. However, it can be used as the last stage in the designed PA due to the reduced output-power levels required by the application.

M1 and M2 are biased in the subthreshold region to achieve high energy efficiency with little dc power. The buffer amplifier that mainly consists of transistor M3 and M4 is self-biased to reduce the second-order harmonic. Capacitors C1, C2, and C3 are dc-block capacitors. L1 is the interstage matching inductor between the

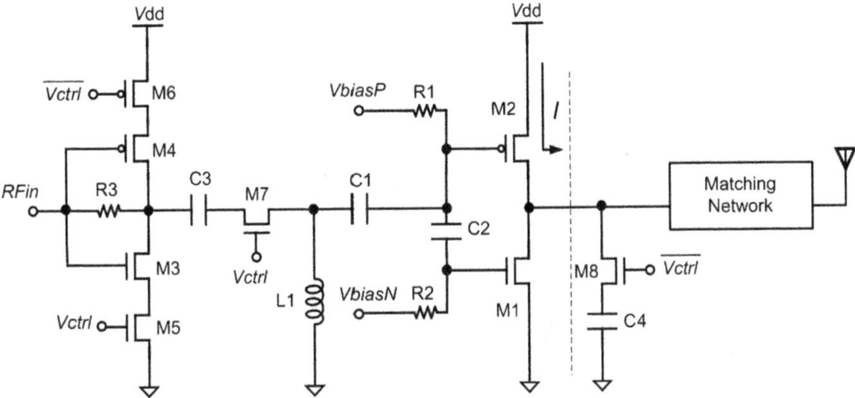

Fig. 8 Schematic of the PA

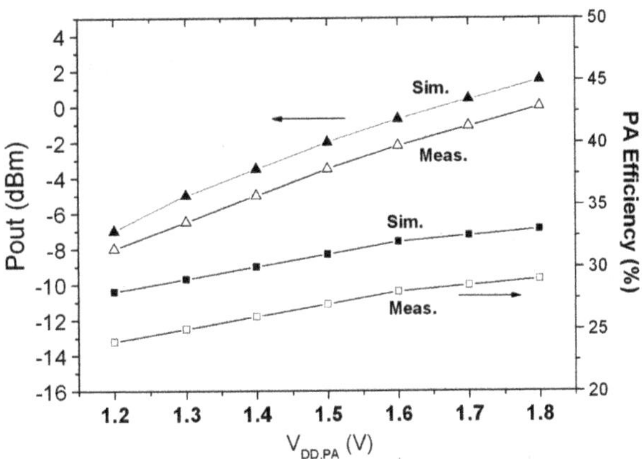

Fig. 9 Output power and efficiency of the PA

buffer and PA. In order to reduce rising time and falling time of the transmitter, complementary switches M5, M6 and M7, M8 are adopted. An off-chip L-type matching network is adopted for its high Q inductor.

The saturation ac current I of the PA is a function of the target output power and the supply voltage. The PA output power is proportional to the square of the current I. And the PA drain efficiency is defined in

$$\text{Efficiency} = \frac{P_{out}}{P_{dc}} = \frac{I_{peak}^2 R_L / 2}{V_{dd} I_{avg}} \propto \frac{I_{avg}}{V_{dd}} \tag{5}$$

Figure 9 shows the simulated and measured PA efficiency and PA output power with a 50 Ω antenna. The measured results show that the PA output power is about 1.5 dB lower than the simulated results, which is mainly due to the parasitical effect of the bonding wire and printed-circuit board (PCB). The energy efficiency of the PA is about 27.8 % under the 1.8-V supply voltage.

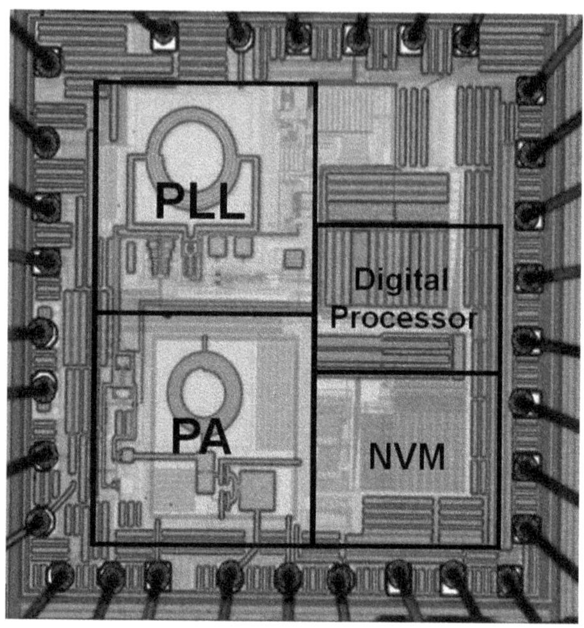

Fig. 10 Die micrograph of the 2.4 GHz transmitter

2.3 Measured Results

The designed transmitter is implemented in the 1P6M 0.18 μm RF CMOS process with a 1.8-V power supply. The transmitter die micrograph is given in Fig. 10. The chip active area is 1.3 mm^2.

The dependence of the VCO output frequency on the presetting signals C and P is shown in Fig. 5. The resolution of the presetting frequency was within 0.6 MHz, as shown in Fig. 7. The mixed-signal VCO covers frequencies ranging from 2.03 to 2.59 GHz. The reference frequency and loop bandwidth of PLL are 1 MHz and 80 kHz, respectively. The measured PLL lock-in time is shown in Fig. 11. The figure shows typical frequency hopping characteristics of the synthesizer at 250 μs. The dependence of the frequency on time shows that the frequency hops rapidly from 2.41 to 2.428 GHz at 250 μs. The lock-in time is less than 3 μs, which is much shorter than that of the conventional PLL.

The lock-in time almost is not dependent on the frequency step, temperature, or process variation. The results demonstrated that the frequency presetting technique could accurately preset the output frequency of VCO and reduce the lock-in time of the PLL. The measured PLL phase noise is −102 dBc/Hz at 1 MHz offset and −117 dBc/Hz at 3 MHz offset, respectively. An ultra-low-power 192-b NVM memory was also implemented. The measured results indicate that the memory can operate well under a wide supply voltage and clock frequency range. Its power consumption is 4 μA (10 μA) at the read (write) rate of 1.3 Mb/s (0.8 kb/s) with the clock frequency of 1 MHz.

The measured output spectrum of the transmitter with OOK modulation is shown in Fig. 12. The spectrum of the modulated signal with a data rate of 4 Mb/s was measured at a carrier frequency of around 2.4 GHz. In the PLL-based mode and

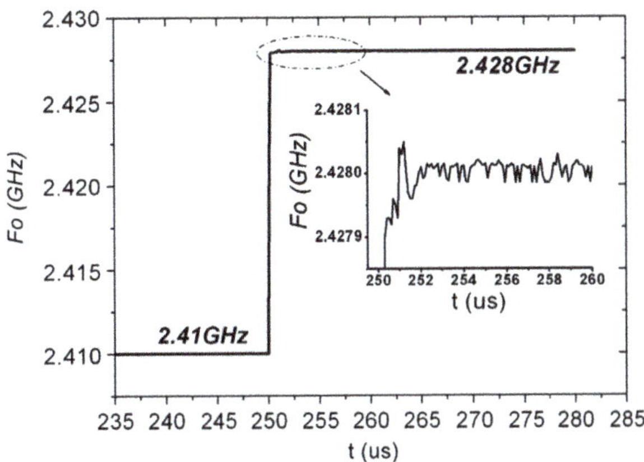

Fig. 11 Measured typical frequency-hopping characteristics of PLL

VCO direct modulation mode, the maximum data rate for OOK modulation could achieve the data rate of 4 Mb/s. The maximum data rate for OOK modulation is limited to the startup time of VCO and PA. The power consumption of the transmitter in the PLL-based mode is larger than in VCO direct modulation mode.

The measured spectrum of the FSK modulation signal for the PLL-based mode is shown in Fig. 13a. The spectrum was measured so that the transmitting data rate was 200 kb/s with 1-MHz frequency deviation. The measured FSK error is 6.88%, which is equivalent to 23 dB SNR. This signal quality is adequate for typical FSK

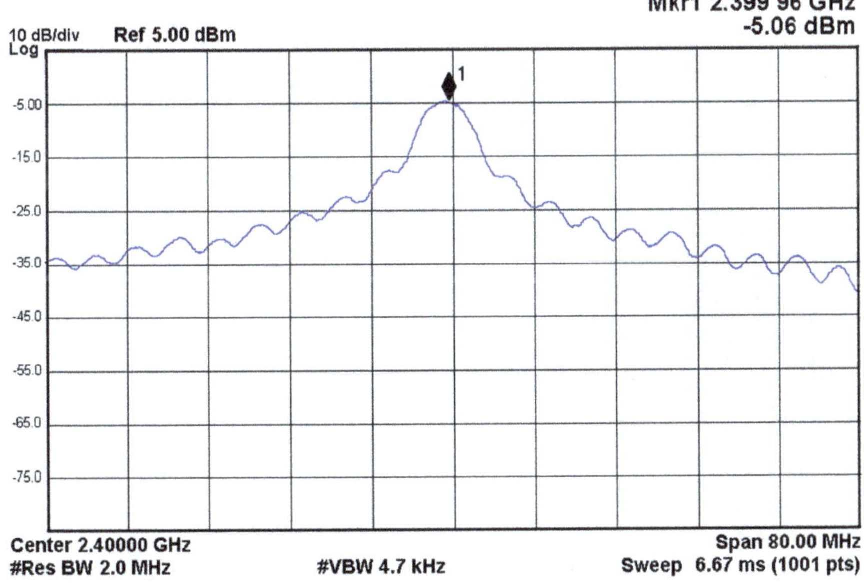

Fig. 12 Spectrum of the OOK modulated signal at 2.4 GHz AT 4 Mb/s

H. Jiang et al.

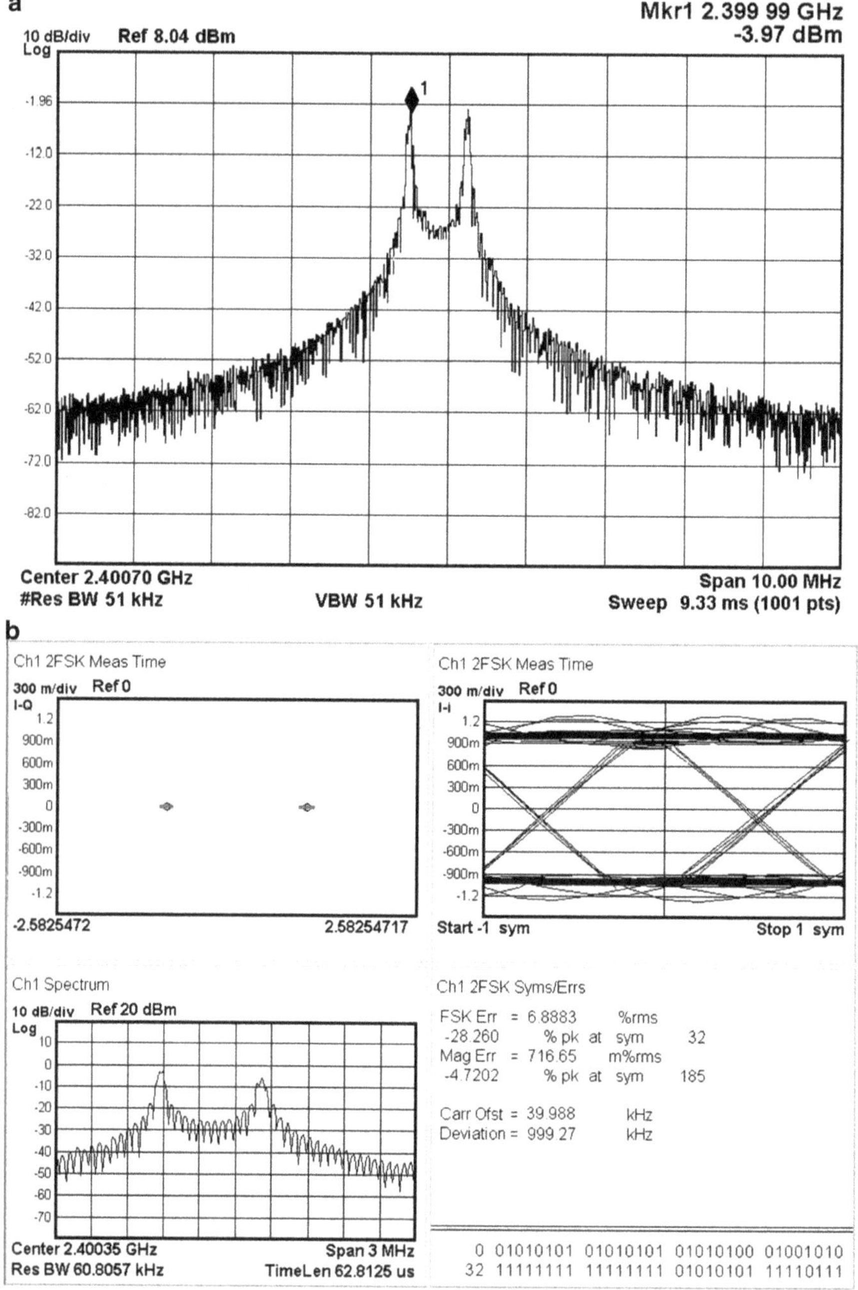

Fig. 13 Measured 200-kb/s FSK modulation characteristics. (**a**) Modulation spectrum. (**b**) FSK error and eye diagram

Table 1 Measured 2.4 GHz transceiver performance

Modulation scheme	OOK/FSK
Process	0.18 μm CMOS
Maximum data rate	4 Mbps(OOK)/2 Mbps(FSK)
PLL phase noise	-102 dBc/Hz@1MHz
PLL lock-in time	3 us(typ.)/1 us(min.)
PA output power	0 dBm
Total TX average current (min.)	2.7 mA(OOK)/5.4 mA(FSK)
VCO	2.3 mA
PLL (include VCO)	3.8 mA
Buffer	1 mA
PA	2 mA
Digital Processor (include NVM)	0.1 mA
FOM	1.2 *nj/bit*·mW (OOK w/o PLL)
	1.6 *nj/bit*·mW (OOK w/ PLL)
	4.8 *nj/bit*·mW (FSK w/o PLL)
	6.2 *nj/bit*·mW (FSK w/ PLL)
Chip core area	1.1×1.2 mm

demodulators to satisfy a bit-error rate (BER) of better than 10^{-3}. In VCO direct modulation mode, the maximum data rate is 2 Mb/s with 10 MHz frequency deviation, and the FSK error is 10.3%. The maximum data rate for FSK modulation is mainly limited to the settling time of PLL, frequency deviation, and system required EVM.

With a 1.8-V supply voltage, the measured minimum average current consumed by the transmitter is 2.7 mA for OOK modulation and 5.4 mA for FSK modulation. A key metric for measuring the energy efficiency of wireless transmitter is the average amount of energy required to transmit a single bit of data. To take into account the transmitter output power, which reflects the efficiency of PA, an *FOM* defined in [16] can be used for efficiency evaluation.

$$\text{FOM} = \frac{PowerConsumption}{P_{out} \times Datarate}(nJ / bit \cdot mW) \tag{6}$$

According to (6), the designed transmitter energy efficiency is 1.2 *nj/bit · mW*/4.8 *nj/bit · mW* for OOK/FSK modulation with the data rate of 4 Mbps/2 Mbps. Table 1 summarizes the measured performance of the transmitter. Although some other designs fabricated in a more advanced technology, such as RF-MEMS technology, have higher efficiency [17, 18], this design has more flexibility in terms of programmability and multi modulation schemes. The implemented transmitter has achieved relatively high energy efficiency while delivering a reasonable output-power level for wireless medical applications.

In summary, this 2.4-GHz energy-efficient transmitter is designed for wireless medical applications. The transmitter combines the VCO direct modulation mode and the PLL-based mode for the optimal energy efficiency. In particular, the data rate of the PLL-based mode can be increased by the frequency presetting technique with low power consumption. The frequency auto-calibration mechanism has been implemented in the design. The calibration parameters and the control signals can be stored in the NVM so that the transmitter can avoid the repetitive calibration process and save the energy in practical applications.

3   400 MHz Transceiver for Implantable Medical Devices

The implantable medical devices (IMD's) usually have very stringent requirements on the form factor. Since IMD's are powered by miniature batteries, the transceiver used for IMD's should have extremely low power consumption. In this subsection, a design example of a 400 MHz transceiver designed for implantable wireless medical devices will be discussed to illustrate some design techniques that can be used for ULP transceivers with the constraints in the implantable medical applications. Note that the 400 MHz band is chosen for with the trade-off between through-body transmission loss and the antenna size.

3.1   Transceiver Architecture

The power consumption is the key issue for this transceiver. The overall structure of the transceiver used in this design in shown in Fig. 14. MSK modulation has been chosen for the transmitter, since MSK is a constant-envelope modulation which can help to alleviate the design requirement on the hub receiver. A high data rate of 3 Mbps is chosen such that the transmitter will in a burst-mode and the average power consumption of the transmitter will be quite low. The receiver takes a 64 kbps OOK modulation for circuit simplification. Since the receiver is turned on very occasionally, the transmitter will consume most of the system energy. Consequently, in this design, the effort of power reduction has been mainly been paid to the transmitter.

The transceiver works at the 400 MHz frequency band. A PLL frequency synthesizer is implemented on-chip to provide the quadrature LO signals for the transmitter and the single phase LO signal for the receiver. The active circuit of a 24 MHz crystal oscillator has been implemented to provide the reference clock for the PLL and the system clock for the digital circuit.

As shown in Fig. 14, the MSK transmitter is composed of a quadrature direct digital synthesizer (DDS) which converts the transmit data into the zero-IF analog MSK baseband signal, and a single side band (SSB) up-converter which converts the IF signal into the 400 MHz frequency band. The DDS consists of a digital MSK modulator and a quadrature DAC pair. The SSB up-converter is mainly a quadrature up-mixer with the preceding low-pass anti-aliasing filter. The OOK receiver is composed of the LNA for RF signal amplification, the down mixer for frequency down shifting, the band-pass filter to suppress the out-of-band disturbance, the cascading programmable gain amplifier (PGA) gain stages for IF signal amplification, the 3-bit ADC for quantization, and the IF DSP that performs the OOK demodulation.

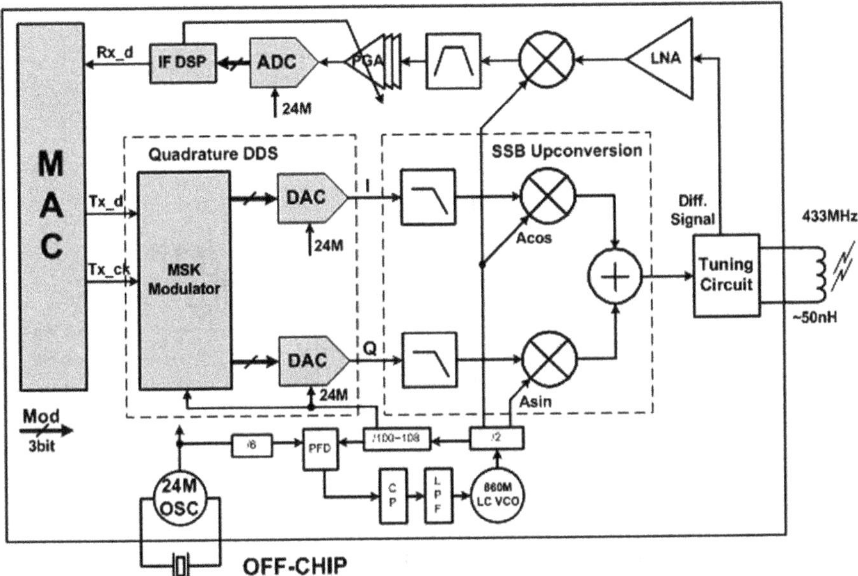

Fig. 14 400 MHz transceiver architecture

In this design, all the power supply voltages are generated by on-chip linear regulators, as the switch regulator using inductors is not a good choice due to the number of external components and the consideration of EMI. Taking this in mind, the effort to lower the circuit power is equivalent to lower the circuit current. For this reason, current reusing techniques have been greatly adopted in the transceiver circuit. The current reusing techniques are mainly applied to the frequency synthesizer and the transmitter.

3.2 ULP Circuit Design Techniques

Figure 15 shows the structure of the frequency synthesizer. The VCO is designed to run at ~800 MHz to generate quadrature phase local oscillation signals. In this synthesizer, the quadratic frequency divider shares the same DC current path as shown in Fig. 15 [19]. A classical PLL locks the VCO frequency. The PLL's divider is programmable, giving the SoC the capability of frequency hopping. The VCO has a coarse tuning circuit which helps to calibrate its center frequency automatically. Both the VCO tuning circuit and the phase frequency detector (PFD) takes the reference frequency from the on-chip 24 MHz crystal oscillator.

The structure of the transmitter is shown in Fig. 16. In this transmitter, the MSK modulator receives the data stream from the main controller, transforms it into zero-IF baseband waveforms, and then sends the parallel baseband data to the DACs.

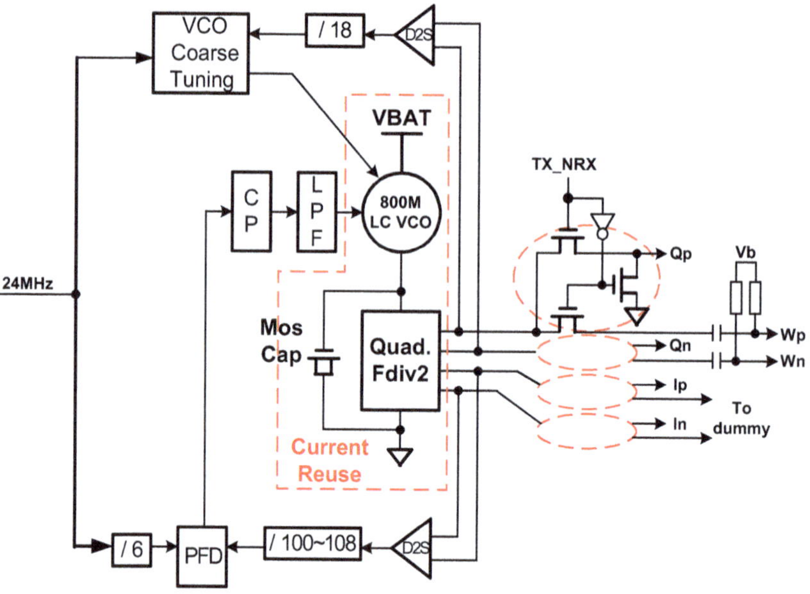

Fig. 15 Quadrature LO Oscillator

There are two key points in this transmitter structures. First, the DC currents of the two DACs are reused by the quadrature mixer (M1 and M2 are used to set the DC current path). Secondly, there is no traditional power amplifier in this transmitter, and a coil which serves as the RF energy emitting component is directly connected to the mixer. This specific structure has been proven effective in the special application environment.

The transceiver uses a simple coil as the antenna. And the transmitter and the receiver share an automatic tuning network. Note that in this specific application, it is really difficult to achieve good antenna matching for the implanted device with a traditional antenna, since the surrounding environment affects the antenna characteristic a lot, and it is quite difficult to characterize antenna's surrounding environment. In this design, the coil antenna can be viewed as an inductor as shown in Fig. 16 and 17. C_{tune} is calibrated automatically so that the RF port achieves resonance. There would be no traditional antenna in this design, and therefore there is no need to do traditional antenna matching for this system.

Actually, in this design a virtual transformer works as the antenna as illustrated in Fig. 17. Inductor L_{emit} actually serves as the primary coil emit, and the human body serves as the secondary coil. Of course, the transformer has very low coupling, which brings the benefit that the transmitter sees very steady load, and this helps to ease the circuit design work. On the other hand, the transmitting loss is quite high due to the low coupling effect, and this can be managed by set the RF ac current level flowing through L_{emit}. Our experience is that if inductor L_{emit} has an RF ac current

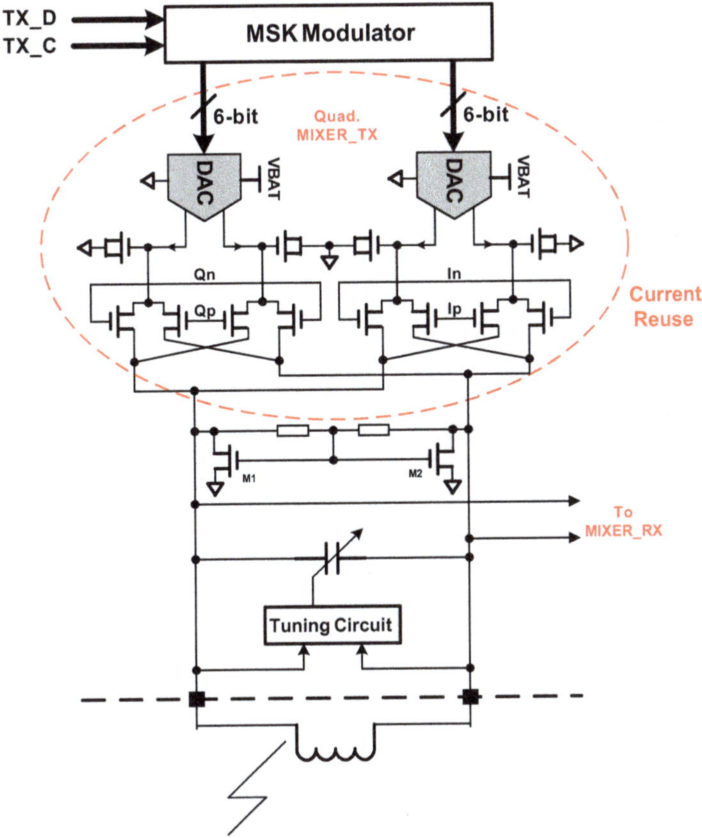

Fig. 16 Transmitter circuit with current reusing

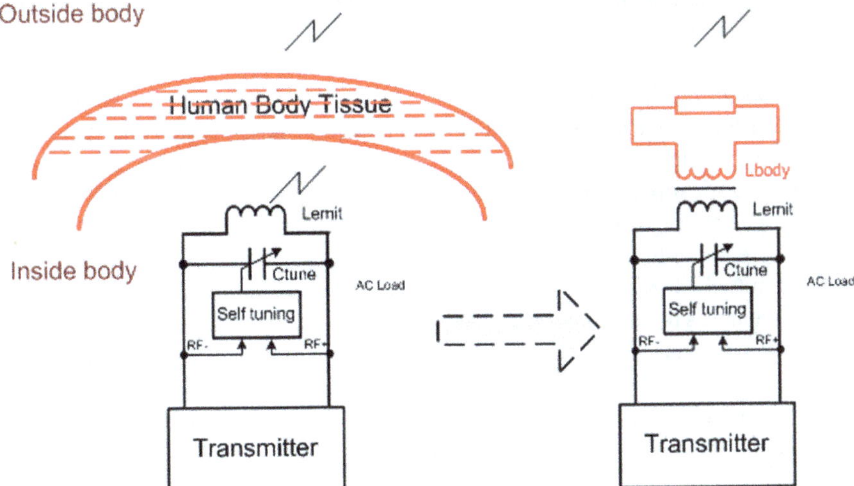

Fig. 17 Virtual transformer for through-body RF signal transmission

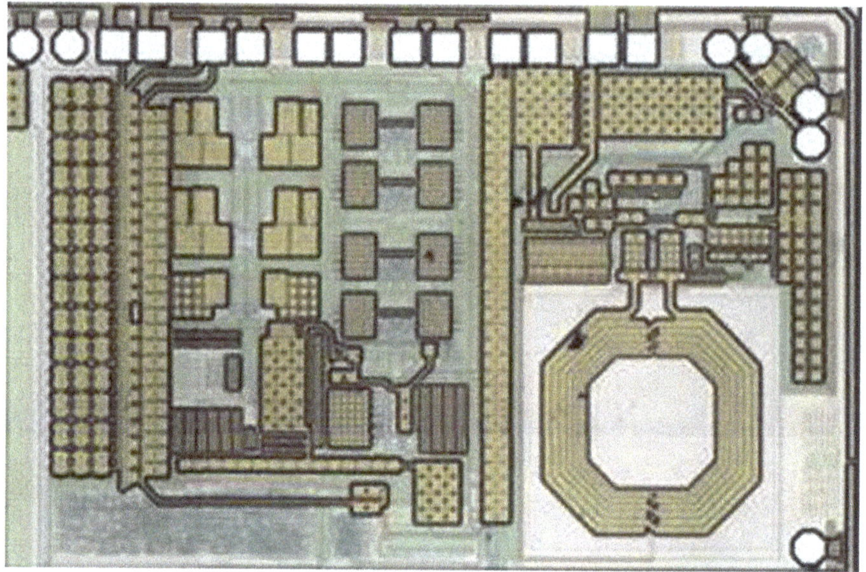

Fig. 18 Microphotograph of 400 MHz transceiver

of 10 mA (peak to peak), the external data recorder antenna can receive >-75dBm RF signal from outside the human body, which is adequate for data demodulation.

3.3 Measurement Results

The ultra-low-power transceiver was fabricated in 0.18-μm 1P6M CMOS process. The prototype chip is shown in Fig. 18. The bit rate of TX circuits is 3 Mbps and the power consumption is 3.9 mW, while and the bit rate of RX circuits is 64 kbps and the power consumption is 12 mW.

To measure the fabricated transceiver, a capsule endoscope prototype with the transceiver PCB is encapsulated and put into a glass beaker (diameter 30 cm) filled with normal saline to emulate the real working environment as shown in Fig. 19.

In the verification, a signal generator is programmed to give an OOK signal (0 dBm transmitting power). The capsule can effectively receive the OOK signal, and interpret the information modulated in the OOK signal.

The emitted MSK signal from the prototype capsule is then measured using a dipole antenna as the external receiving device. The received RF signal power spectrum with a data rate of 3 Mbps is measured as shown in Fig. 20a. The received signal power by the external device can reach ~ -55 dBm, which is far adequate for external receiving if the external data logger has a receiving sensitivity of -90 dBm. The demodulated MSK signal (analyzed using a NI virtual instrument) is shown in

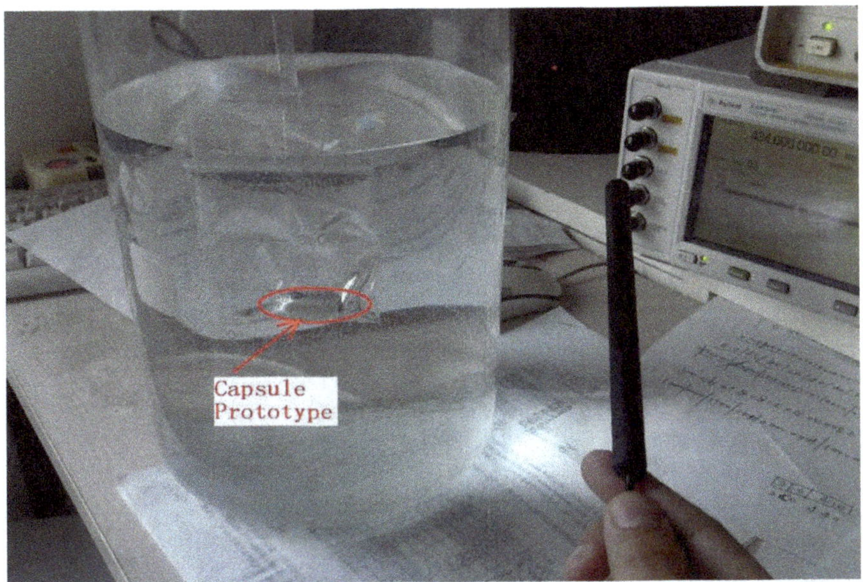

Fig. 19 Capsule endoscope prototype with designed 400 MHz transceiver in emulated environment

Fig. 20b with the received IF signals (I & Q) (left-down), the eye-diagram (right-up), and the phase transition diagram (right-down).

The performance of the implemented ultra-low-power 400 MHz transceiver is summarized in Table 2. The transmitter has a transmission efficiency of 1.3 nJ/bit which makes it quite suitable for the implantable wireless medical applications.

4   A Multiband Transceiver Front End for WBAN Hubs

Many WBAN wireless transceivers work in 433, 868 or 915 MHz, 2.45 GHz license-free industrial, scientific and medical (ISM) frequency band, but with different modulation schemes and different signal bandwidths [20, 21]. To design a WBAN hub that supports various medical applications with one single transceiver chip, it is necessary to implement a reconfigurable transceiver compatible with various SID node transceivers. In this subsection, a reconfigurable multi-mode multi-band transceiver front end for sub-1 GHz short-range wireless applications is discussed. The low-IF receiver with 3 MHz IF carrier frequency and the direct-conversion transmitter provide reconfigurable communication bandwidths from 250 kHz to 2 MHz. An integrated multi-band phase-locked loop frequency synthesizer is utilized to provide the quadrature LO signals for the transceiver. The transceiver supports a highest data rate of 3 Mbps for MSK modulation. The operation frequency can be reconfigured to various ISM frequency bands at 433, 868 or 915 MHz,

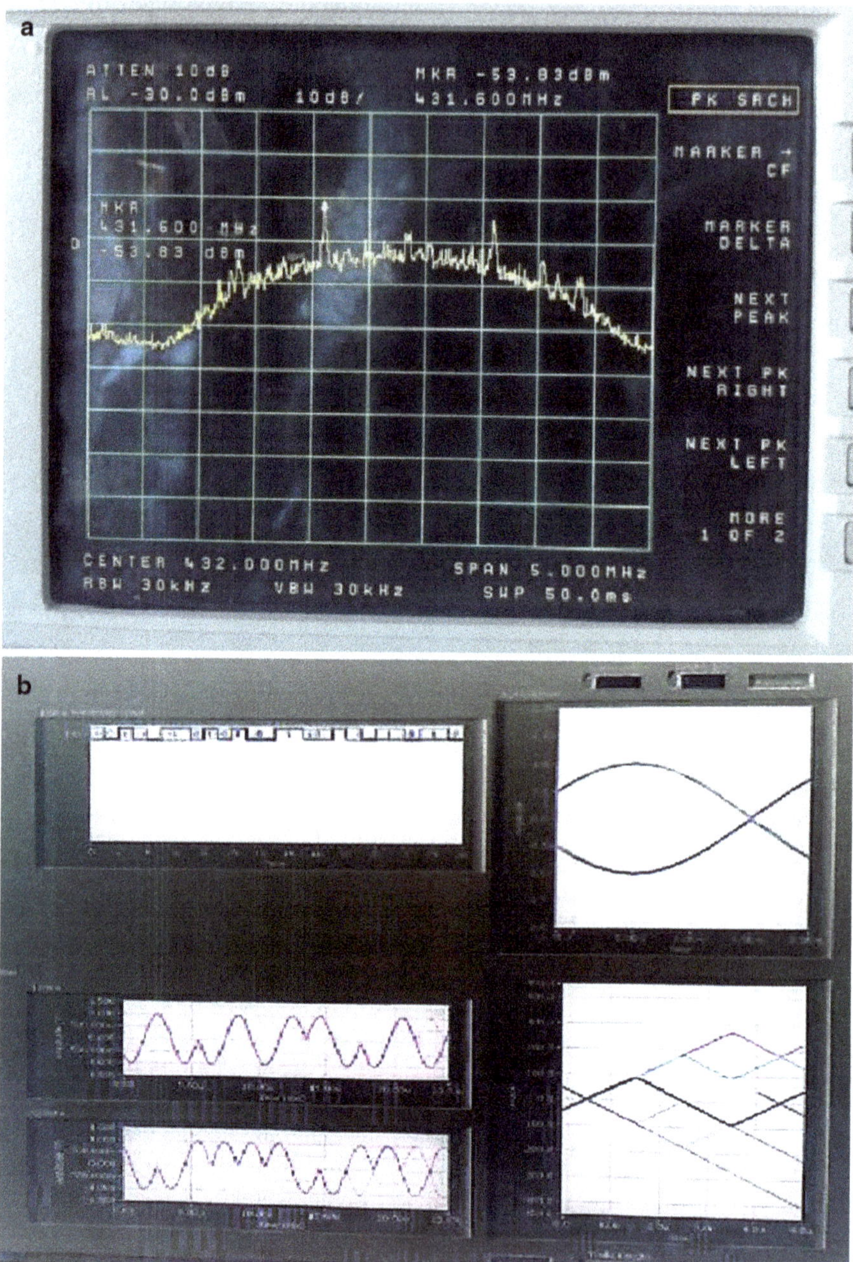

Fig. 20 MSK signal output from the SoC. **a** Transmitter output power spectrum, **b** Demodulated MSK signal

Table 2 Performance of 400 MHz IMD Transceiver

	Type of RF link	Bi-directional
TX	Bit rate	3 Mbps
	Modulation type	MSK
	Power consumption	3.9 mW
RX	Bit rate	64 kbps
	Modulation type	OOK
	Power consumption	12 mW
Technology		0.18 μm CMOS

402~405 MHz MICS band in North America [22] and 407–425 MHz in-hospital medical band in China [23], namely 400 and 900 MHz band. The transceiver has been implemented in 0.18 μm CMOS process. The measured results verify the feasibility of the reconfigurable multi-band transceiver.

4.1   Transceiver Architecture and System Design

A block diagram of the reconfigurable multi-band transceiver front end is illustrated in Fig. 21. It consists of the RF front-end and the IF analog signal processing block. The on-chip receiver outputs the I and Q analog signals directly, and the I and Q analog signals input to the on chip transmitter directly as well. A digital baseband circuit can connect to this chip through off-chip analog to digital converters (ADCs) and digital to analog converters (DACs) to form the complete receive (RX) and transmit (TX) path.

Some necessary auxiliary function blocks such as the bandgap reference, the low-dropout voltage regulator (LDO) and the crystal oscillator are also integrated on the chip. Only a few external components required including a passive transformer to match the RF ports to an antenna, a crystal and a bias resistor are needed.

Receiver Architecture

The receiver utilizes the low-IF architecture to achieve an area-efficient monolithic integration as well as low power consumption and high performance in a CMOS technology [24]. The receiver consists of a low noise amplifier (LNA), a down-converter and IF analog signal processing blocks. The IF blocks include the multi-stage programmable gain amplifiers (PGAs) for signal gain adjusting, and the reconfigurable complex band-pass filter (CBPF) for image rejection and channel filtering. Additionally, a logarithmic received signal strength indicator (RSSI) block is implemented for automatic gain control (AGC), and an on-chip automatic frequency tuning block is introduced for accurately bandwidth control.

A wideband LNA with 2.5~3 dB NF is chosen to support 300 MHz to 1 GHz band. An IF-frequency of 3 MHz is selected based on a careful tradeoff between achievable data rate and power consumption while the chip area and selectivity requirements are also taken into consideration. Channel filtering and image rejection are implemented using a third-order active poly-phase CBPF with a reconfigurable

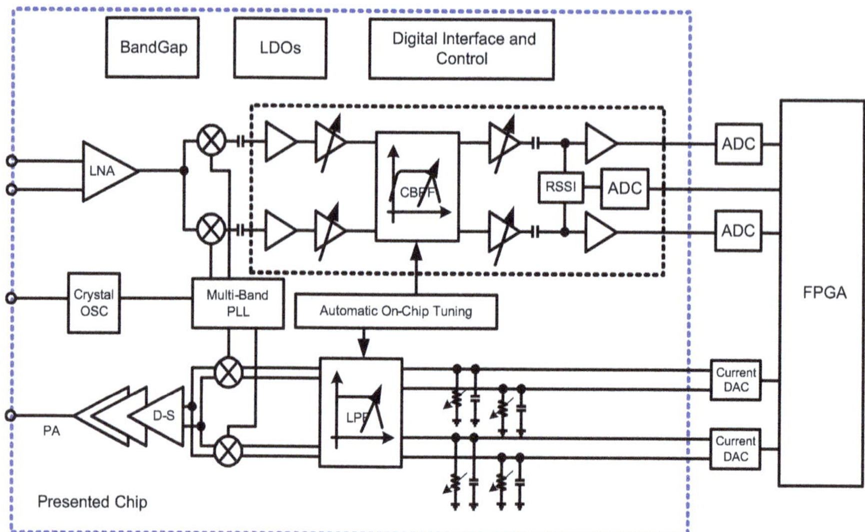

Fig. 21 The block diagram of the presented reconfigurable multi-band transceiver

bandwidth from 250 kHz to 2 MHz. The filter has a Bessel response with its lower Q-factor and flatter group delay to satisfy requirements of various modulation schemes, especially the MSK modulation. This on-chip analog channel filter provides necessary suppressing to image signal and large blocking signal and acts as the anti-aliasing filter for the off-chip ADCs. Strict channel selecting is done by the high order FIR digital filter on the FPGA after the ADCs.

The receiver should totally provide a maximum gain of about 75 dB according to the system level specification [25]. Since the power consumption of IF blocks is restricted, one third of this gain is provided by the RF front-end with the trade-off between the global noise and the non-linear performance. The AGC controls not just the IF gain, but also the LNA gain settings to expand the gain tuning range by about 20 dB variation. The PGAs provide $15 \sim 51$ dB gain with 3 dB programmable. Although usually it is a good choice to place the CBPF in front of the PGAs to suppress the out-of-band interferences and to alleviate the linearity requirement on the PGAs, it is quite difficult to design a CBPF that satisfies the noise requirement at the acceptable low power level. As a trade-off, a noise optimized fixed gain amplifier (FGA) is introduced as the first stage of the analog IF circuit. Following the FGA is a linearity optimized PGA stage, the CBPF, and two other FGA stages, as shown in Fig. 21.

Transmitter Architecture

The transmitter utilizes the direct-conversion architecture to achieve low power consumption. It has a reconfigurable signal bandwidth from 250 kHz to 2 MHz and consists of a three-order active RC low-pass filter (LPF), an up-converter, a

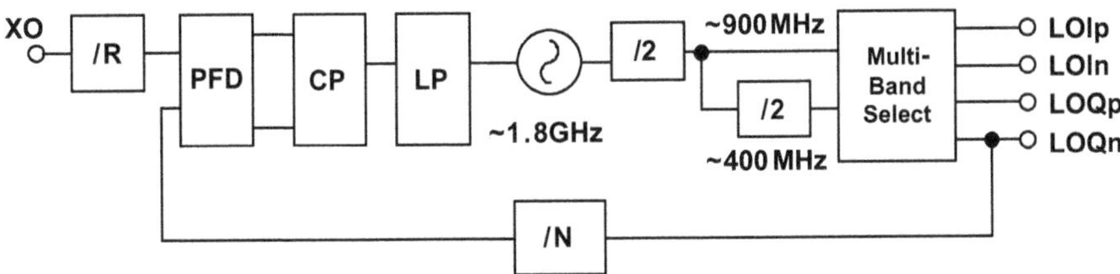

Fig. 22 Simplified diagram of PLL frequency synthesizer

differential-to-single-ended converter (D-S) and a power amplifier (PA). The first-order RC networks with programmable impedance are used as the loads of the external current DACs and convert the input current signals into the voltage signals with programmable gain. The power amplifier has a single-ended output and covers the 300 MHz to 1 GHz frequency band.

Frequency Synthesizer Architecture

A PLL frequency synthesizer is integrated to provide the multi-band quadrature LO signals with reconfigurable channel spacing and high spectral purity.

Figure 22 shows the simplified diagram of the multi-band PLL frequency synthesizer. A LC VCO is used to generate the high frequency signal. Two frequency dividers following the VCO generate the multi-band quadrature LO signals. One of the signals is selected by the multi-band selection block, and then sent to the mixer in the RX or the TX path.

4.2 Key Circuit Descriptions

LNA

Figure 23 shows the simplified schematic of the wideband LNA. The NMOS input pair M1/M2 provides the amplifier's transconductance (G_m). The PMOS M3/M4 are introduced as a parallel input pair to enhance the overall Gm. Particularly, M3 and M4 conduct most part of the tail current I_B, and larger load resistors can be used to boost the LNA gain without imposing any higher DC voltage headroom so that the receiver can use a low supply voltage to maintain the power consumption low. The LNA uses the active shunt feedback for wideband input to match with the cascode transistors M5/M6 and source follower transistors M7/M8, forming the active feedback path with a wideband resistance characteristic as given by

$$Z_{in} = \frac{2}{g_{m7,8}(1+G_{m,in}Z_C)} \tag{7}$$

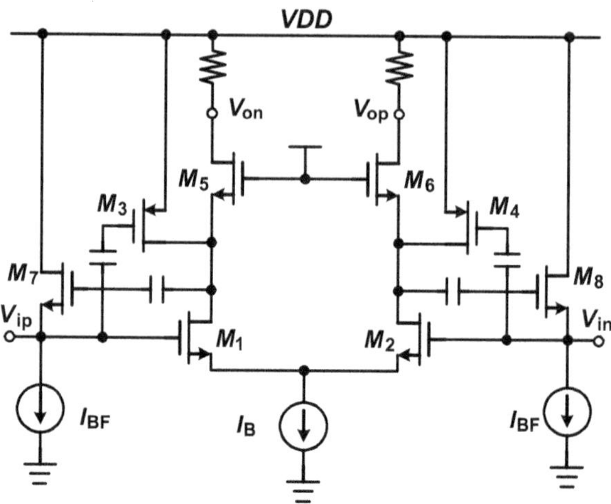

Fig. 23 Schematic of the wideband LNA

where $g_{m7,8}$ stands for transconductance of M7/M8, $G_{m,in}$ is the total G_m of the input transistors M1/M2 and M3/M4, and Z_c is the impedance of the cascode node which is mainly determined by the node parasitic capacitor and cascode G_m. The overall noise factor F_n is given by:

$$F_n = 1 + \frac{\gamma}{\alpha}\left(\frac{1}{1 + G_{m,in}/G_{m5,6}} + \frac{1}{R_S G_{m,in}}\right) + \frac{1}{R_S R_L G_{m,in}^2} \tag{8}$$

where γ is the excess channel thermal noise coefficient, and α approximately equals to 1. R_S and R_L stand for the source and load resistance, respectively. Equations (7) and (8) show that the proposed LNA architecture can provide a flat input matching impedance and an optimized noise figure (NF) simultaneously over the wide frequency band, and avoid the confliction between the impedance matching and NF performance usually seen in the common gate LNA [26].

Complex Bandpass Filter

Figure 24 shows the designed 3rd-order Bessel active-RC CBPF. It is realized by adding inter-connected resistors between I and Q path of the prototype active-RC LPF.

The op-amp used in the filter is the two-stage op-amp with Miller compensation as shown in Fig. 25. A common-mode feedback loop (CMFB) is adopted to maintain the output common-mode voltage level. There is a start-up circuit within the dashed line in Fig. 25. It is employed here to avoid the dead situation introduced by the interconnected resistors, in which the CMFB loop cannot work properly and the filter works at abnormal DC operating points. In that case, the gate voltage of M4/M5 is extremely low and the last stage output common-mode voltage is as high as

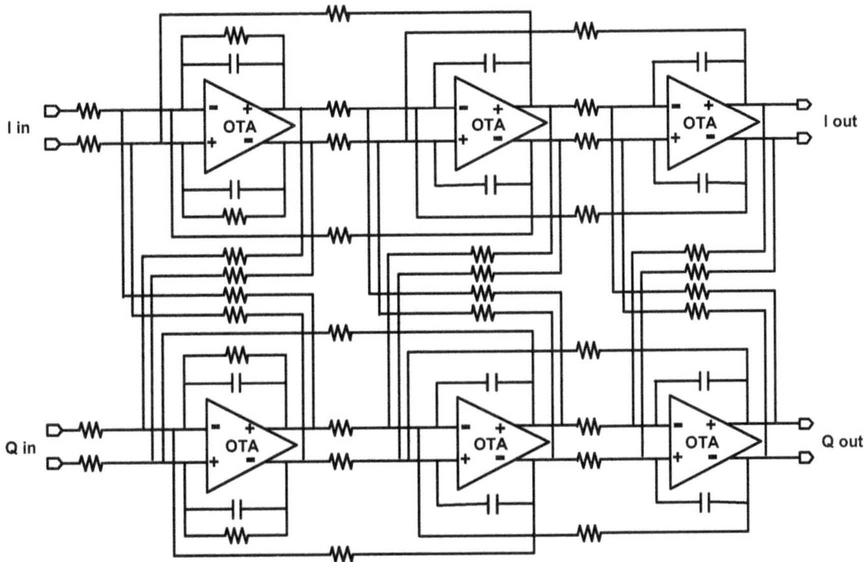

Fig. 24 Third-order Bessel active-RC complex band-pass filter

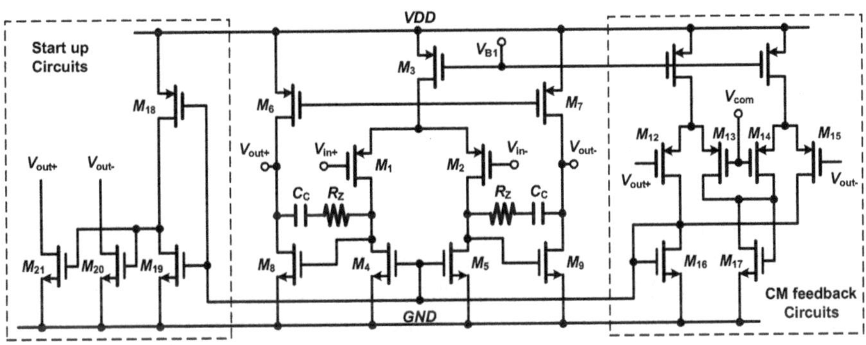

Fig. 25 Two stage op-amp with Miller compensation and start-up circuit

the supply power. An inverter formed by M18 and M19 is employed to break the abnormal stable operating points. The inverter's input is the gate voltage of M4/M5. When it is very low, the inverter output controls M20 and M21 pulling-down op-amp's output and forces the CMFB circuit to restore its proper operation. The W/L of M18 and M19 is carefully designed to avoid affecting the amplifier core circuit when the CMFB loop works normally.

To overcome the problems caused by inaccurate R/C components due to the process and temperature variation [27, 28], the automatic frequency tuning circuit is designed as shows in Fig. 26. V_{BG}/R current is used to charge capacitor array

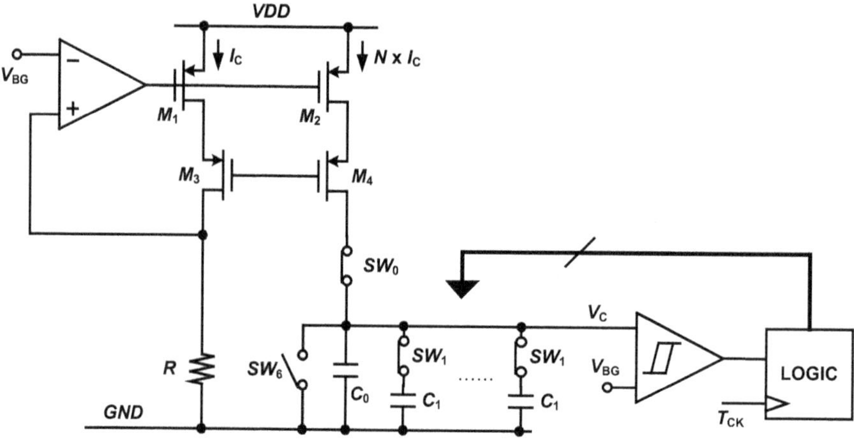

Fig. 26 Automatic frequency tuning circuit

integrator. *RC* time constant is then judged and adjusted by comparing the integrating period to a reference clock.

PGAs

The first stage PGA demands high linearity and low noise performance, since it is located before the CBPF. A negative feedback closed-loop structure is chosen to meet the linearity requirement while its noise performance is acceptable. The variable gains, range from −6 to 18 dB with a step of 3 dB, are realized by switching the resistors in the feedback path.

The requirements for linearity performance are not rigid in the remaining PGA stages, because the CBPF has suppressed the out-of-channel interferences. The open-loop source degeneration amplifiers are employed to decrease power consumption.

RSSI

The successive-detection architecture [29] is used in the receiver to realize the logarithmic-linear form as shown Fig. 27. In this circuit, DC offset must be eliminated, because even a tiny input referred offset voltage would saturate the internal nodes of the amplifier chain. By using the forward DC coupling method as shown in Fig. 28 [30], the DC offset and low frequency noise, extracted by the RC low-pass network, are subtracted at the input stage of the amplifier. A replica biasing for constant-gain amplifier [31] is used; its gain is determined only by the ratio of device dimensions, so as to avoid the second-order effect due to the temperature and process variations.

A 7 bits successive approximation (SAR) ADC is integrated on chip to quantize the RSSI outputs.

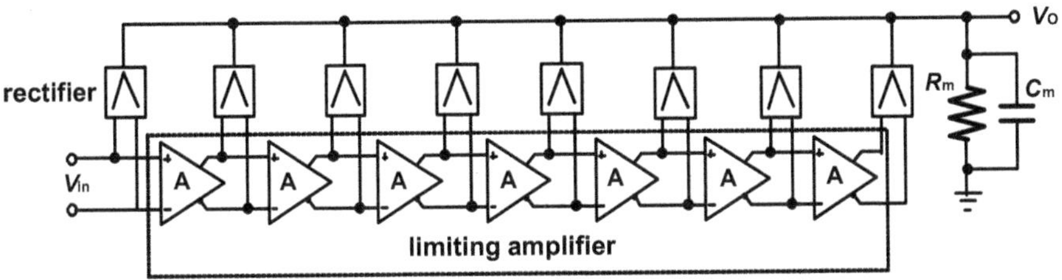

Fig. 27 Block diagram of RSSI

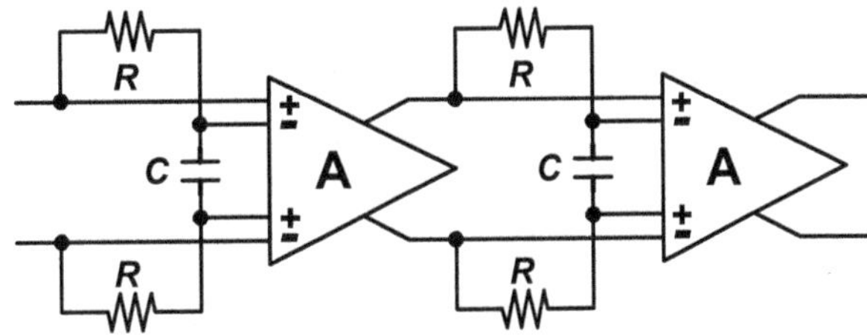

Fig. 28 Forward DC
coupling method used in
RSSI

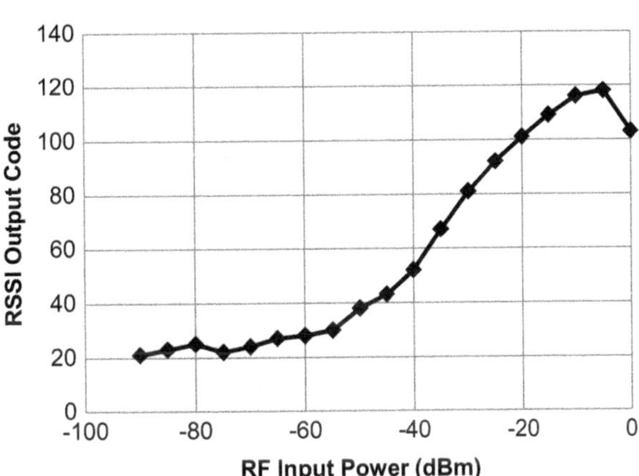

Fig. 29 Measured RSSI
linearity performance

Figure 29 shows the measured RSSI function curve. It is measured when the CBPF bandwidth is set to 2 MHz and the PGA gain is set to the lowest level. In the figure, the horizontal axis is the input power of the first stage FGA in the analog IF block, and the vertical axis is the digital output code of the 7 bit SAR ADC. The linear range of the RSSI circuit is from -60 to -10 dBm referred to the input port of the analog IF chain.

Power Amplifier

Figure 30 shows the schematic of the power amplifier. The differential inputs from the up-converter are converted into the single-ended signal by two common-source

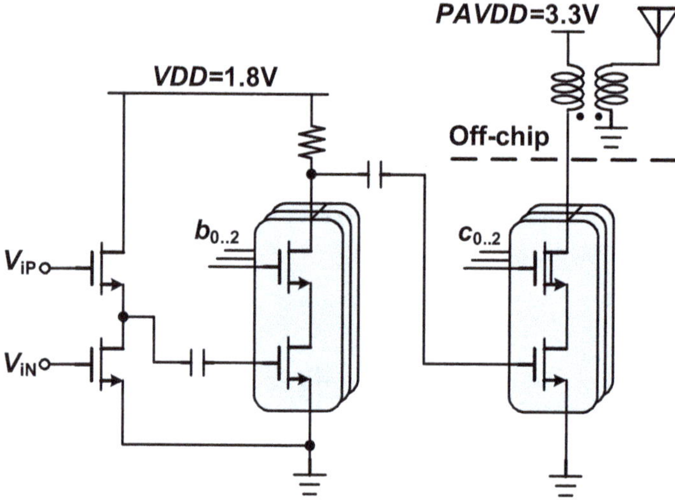

Fig. 30 Simplified schematic of PA

Fig. 31 Measured output power versus the input power of the transmitter

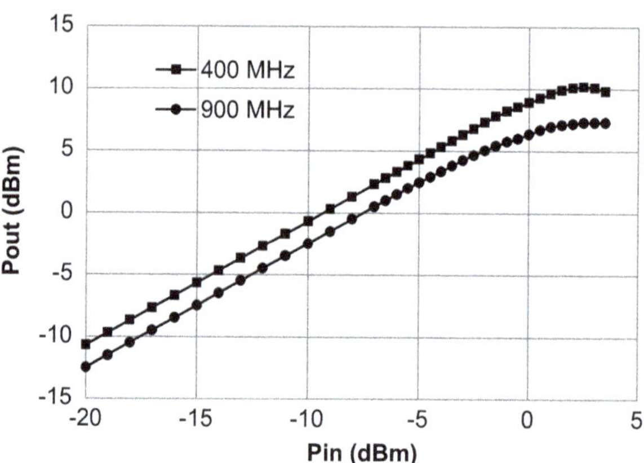

common-drain combined transistors and then amplified by the PA buffer with the wideband load. The last stage of the PA uses 3.3 V power supply and thick-gate/thin-gate transistor combination. The thick-gate transistor has a higher voltage tolerance and improves the output voltage swing while the thin-gate transistor provides higher gain. The electrical characteristics of on-chip part of the PA show a wideband response, so the multi-band operation can be implemented by utilizing the wideband off-chip transformer. The PA provides multi output power levels which are controlled by b_i and c_i ($i=0, 1, 2$).

Figure 31 gives the measured output power versus the PA input power when it is configured in 400 or 900 MHz frequency band and 500 kHz signal bandwidth. The PA can provide a 10.2 dBm maximum output power level for 50 Ω load at 400 MHz, and provide a 7.3 dBm maximum output power at 900 MHz.

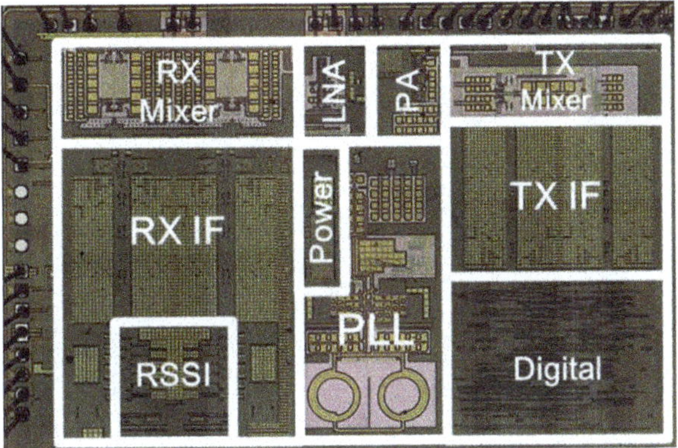

Fig. 32 Microphotograph of the multi-band transceiver front end

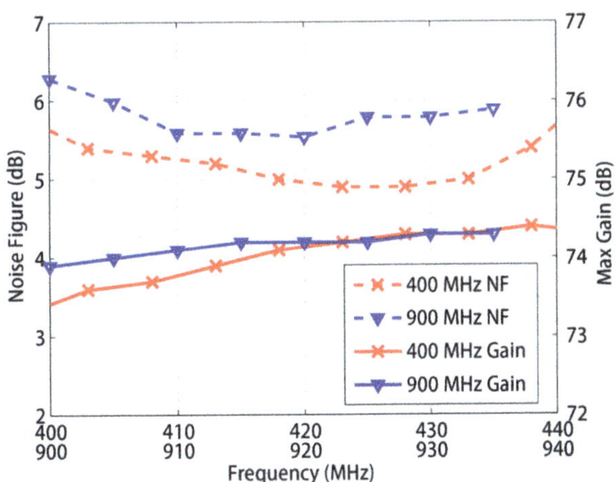

Fig. 33 Measured gain and NF versus frequency of the receiver

4.3 Implementation Results

The reconfigurable multi-band transceiver front end has been implemented in 0.18 μm CMOS process. The supply voltage of the core circuits is 1.8 V except that the PA's output stage uses a 3.3 V power supply. Figure 32 shows the die microphotograph. The active die area is 3.8×2.6 mm.

Figure 33 gives the measured receiver gain and NF versus operation frequency when configured in 400 or 900 MHz frequency band with the highest gain setting. The 400 MHz receiver provides 74.3 dB gain with a 4.9 dB NF, and the 900 MHz receiver provides 74.2 dB gain with a 5.5 dB NF.

Figure 34 shows the two-tone test results of the receiver front end. In this test, 904.7 and 905.3 MHz RF signals are sent to the LNA input node while the PLL is locked at 902 MHz. The output signal of the first stage FGA is measured and used to calculate the front end *IIP3*. The calculated *IIP3* of the 900 MHz band is -18.2 dBm as showed in Fig. 34. The 400 MHz band has a little lower IIP3 of -19.6 dBm.

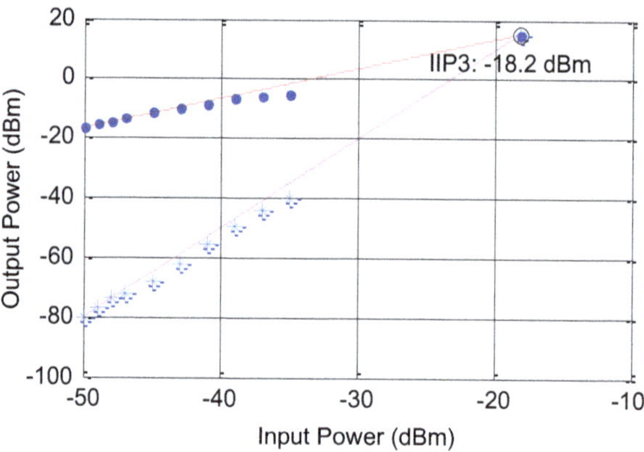

Fig. 34 Measured IIP3 of the receiver front-end in 900 MHz band

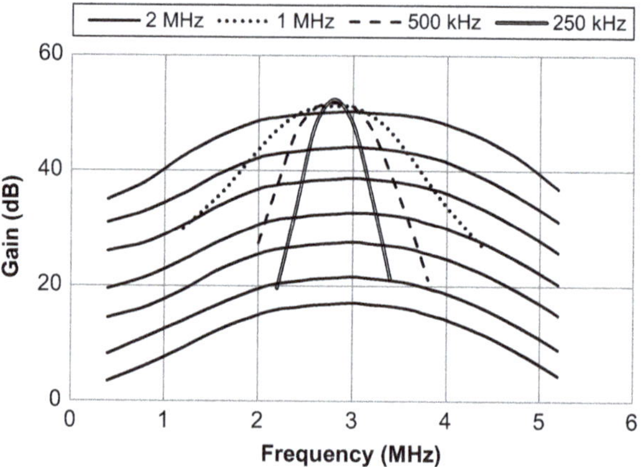

Fig. 35 Measured frequency response curves of the RX analog IF blocks with different signal bandwidth and gain settings

Figure 35 shows the measured frequency response (FR) curves of the receiver analog IF blocks. FRs versus various gain settings in the 2 MHz CBPF bandwidth are drawn in solid line. As seen in the figure, the IF chain achieves the gain range of about 34 dB which is quite close to the targeted value of 36 dB. The reconfigurable bandwidth of the CBPF is also measured at the PGA's maximum gain.

Figure 36 gives the measured phase noise of the PLL frequency synthesizer. The phase noise with the 451 MHz carrier frequency is -107 dBc/Hz @ 100 kHz offset. The loop bandwidth is set to 30 kHz when the measurement is performed. The in-band phase noise is -73 dBc/Hz.

Table 3 summaries the performance of the reconfigurable multiband transceiver.

In summary, the reconfigurable multi-band transceiver front end can be used for low power short-range wireless medical applications. The low-IF receiver with 3 MHz IF carrier frequency and the direct-conversion transmitter supports reconfigurable signal bandwidths from 250 kHz to 2 MHz. The measured results verify the feasibility of the reconfigurable multi-band transceiver.

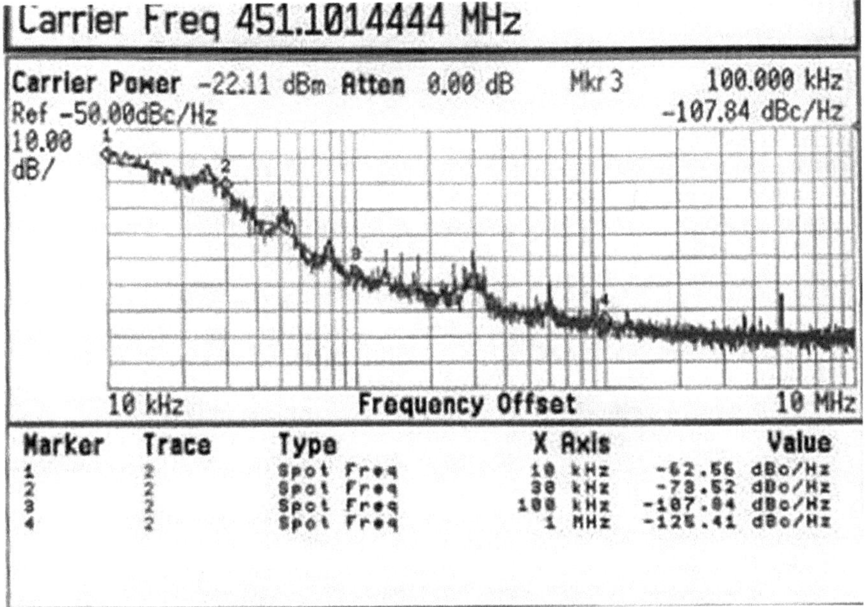

Fig. 36 Measured phase noise of the frequency synthesizer

Table 3 Multiband transceiver front end performance summary

Tuning range (MHz)	310–470		
Channel bandwidth (kHz)		250–2,000	
RX	Noise Fig. (dB)	4.9	5.9
	IIP3 (dBm)	−19.6	−18.2
	Power (PLL+RX) (mW)	32.9	35.6
	Power effciency (nJ/bit)	11.0	11.9
TX	Max. output power (dBm)	10.2	7.3
	Power (PLL+TX) (mW)	47.4	50.1
PLL	Phase noise@100 kHz (dBc/Hz)	−107	−92.7
Technology (um)		0.18	
Die area (mm²)		9.88	

References

1. J. Yoo, L. Yan, D. El-Damak, M. Bin Altaf, A. Shoeb, H.-J. Yoo, et al., "An 8-channel scalable EEG acquisition SoC with fully integrated patient-specific seizure classification and recording processor," in 2012 IEEE International Solid-State Circuits Conference, ISSCC, 2012, pp. 292–294.
2. H. Jiang, F. Li, X. Chen, Y. Ning, X. Zhang, B. Zhang, et al., "A SoC with 3.9 mW 3 Mbps UHF transmitter and 240 uW MCU for capsule endoscope with bidirectional communication," in 2010 IEEE Asian Solid State Circuits Conference, A-SSCC, 2010, pp. 1–4.
3. K. Yang, H. Jiang, J. Dong, C. Zhang, and Z. Wang, "An adaptive real-time method for fetal heart rate extraction based on phonocardiography," in 2012 IEEE Biomedical Circuits and Systems Conference, BioCAS, 2012, pp. 356–359.

4. H. Chen, X. Zhang, M. Liu, W. Hao, C. Jia, H. Jiang, et al., "Low-power circuits for the wireless ligament balance measurement system in TKA," Analog Integrated Circuits and Signal Processing, vol. 72, no. 2, pp. 293–302, Aug 2012.

5. S. Mai, Z. Wang, C. Zhang, and Z. Wang, "A wirelessly programmable chip for multi-channel neural stimulation," in 2012 Annual International Conference of the IEEE Engineering in Medicine and Biology Society, EMBC, 2012, pp. 6595–6599.

6. P. G. Jacobs, J. E. Youssef, J. R. Castle, J. M. Engle, D. L. Branigan, P. Johnson, et al., "Development of a fully automated closed loop artificial pancreas control system with dual pump delivery of insulin and glucagon," in 2011 Annual International Conference of the IEEE Engineering in Medicine and Biology Society, EMBC, 2011, pp. 397–400.

7. G. Z. Yang, Body Sensor Networks, Springer, London, 2006.

8. IEEE Standard for Local and Metropolitan Area Networks—Part 15.6: Wireless Body Area Networks, IEEE 802 LAN/MAN Standards Committee, 2012. [Online]. Available: www.ieee.org

9. B. Latré, B. Braem, I. Moerman, C. Blondia, and P. Demeester,"A survey on wireless body area networks," Wireless Networks, vol. 17, no. 1, pp. 1–18, 2010

10. IEEE Standard for Information Technology—Part 15.4: Wireless Medium Access Control (MAC) and Physical Layer (PHY) Specifications for Low-Rate Wireless Personal Area Networks (LR-WPANs), IEEE 802 LAN/MAN Standards Committee, 2006. [Online]. Available: www.ieee.org

11. X. F. Kuang, and N. J. Wu, "A fast-settling monolithic PLL frequency synthesizer with direct frequency presetting," in ISSCC Dig. Tech. Papers, Feb. 2006, pp. 204–205

12. P. Feng, Y. Li, and N. J. Wu, "An ultra low power non-volatile memory in standard CMOS process for passive RFID Tags," in Proc. IEEE Custom Integrated Circuit Conf., Sep. 2009, pp. 713–716

13. E. Hegazi, H. Sjoland, and A. A. Abidi, "A filtering technique to lower LC oscillator phase noise," IEEE J. Solid-State Circuits, vol. 36, no. 12, pp. 1921–1930, Dec. 2001

14. T.-H. Lin, and Y.-J. Lai, "An agile VCO frequency calibration technique for a 10-GHz CMOS PLL," IEEE J. Solid-State Circuits, vol. 42, no. 2, pp. 340–349, Feb. 2007

15. C. Yoo, and Q. Huang, "A common-gate switched 0.9-W class-E power amplifier with 41 % PAE in 0.25-um CMOS," IEEE J. Solid-State Circuits, vol. 36, no. 5, pp. 823–830, May 2001.

16. M. K. Raja, and Y. P. Xu, "A 52 pJ/bit OOK transmitter with adaptable data rate," in Proc. IEEE Asian Solid-State Circuits Conf., Nov. 2008, pp. 341–344.

17. D. C. Daly, and A. P. Chandrakasan, "An energy-efficient OOK transceiver for wireless sensor networks," IEEE J. Solid-State Circuits, vol. 42, no. 5, pp. 1003–1011, May 2007.

18. B. Otis, Y. H. Chee, and J. Rabaey, "A 400 uW-RX, 1.6 mW-TX superregenerative transceiver for wireless sensor networks," in Proc. ISSCC Dig. Tech. Papers, Feb. 2006, pp. 396–397.

19. K. Park, C. Jeong, J. Park, J. Lee, J. Jo, and C. Yoo, "Current reusing VCO and divide-by-two frequency divider for quadrature LO generation," IEEE Microwave & Wireless Comp. Let., vol. 18, no. 6, pp. 413–415, Jun. 2008

20. V. Peiris, C. Arm, S. Bories, et al., A 1 V 433/868 MHz 25 kb/s-FSK 2 kb/s-OOK RF transceiver SoC in standard digital 0.18 um CMOS, IEEE ISSCC Dige Tec Papers, 2005, 48:258.

21. Li Guofeng, Wu Nanjian, A low power flexible PGA for software defined radio systems, Journal of Semiconductors, 2012, 33(5):055006.

22. FCC Rules and Regulations, MICS Band Plan, http://www.fcc.gov

23. China Radio Management, Technology regulations for micro-power short-range wireless device, http://www.miit.gov.cn

24. J. Crols, and M. Steyaert, Low-IF topologies for high-performance analog front ends of fully integrated receivers, IEEE Trans Circ Syst II, 1998, 45(3):269.

25. M. El-Nozahi, E. Sánchez-Sinencio, Power-Aware Multiband–Multistandard CMOS Receiver System-Level Budgeting, IEEE Trans Circ Syst II, 2009, 56(7):570.

26. F. Belmas, F. Hameau, J. Foumier, A low power inductorless LNA with double Gm enhancement in 130 nm CMOS, IEEE J Solid-State Circuits, 2012, 47(5):1094.

27. H. Majima, H. Ishikuro, K. Agawa, et al., A 1.2-V CMOS complex bandpass filter with a tunable center frequency, IEEE ESSCIRC Proceeding, 2005, 30:327.

28. Wu Ende, Yao Jinke, Wang Zhihua, 9 MHz Active-RC Filter with On-Chip Digital Controlled Frequency Tuning, Chinese Journal Of Semiconductors, 2005, 26(6):1250.
29. C. Yang, A. Mason, Precise RSSI with High Process Variation Tolerance, IEEE ISCAS, 2007, 2870.
30. A. Richards, DC Blocking Amplifier, 1991, US, 5073760.
31. G. Palmisano, R. Salerno, A replica biasing for constant-gain CMOS open-loop amplifiers, IEEE ISCAS, 1998, 2:363.

Low Power Energy Metering Chip

Kun Yang, Shupeng Zhong, Quan Kong, Changyou Men
and Nianxiong Nick Tan

1 Introduction

Energy meters are usually directly floating on the live wire due to the direct current sensing (e.g. shunt resistors) and/or voltage sensing (resistance dividers). For safety's viewpoint, the meters have to be isolated to protect human beings from 220 V power line. The isolation is usually accomplished by opto-couplers. There are two ways of isolations. One is to isolate the whole meter by floating the metering chip and MCU (Micro Control Unit) on the live wire and using opto-couplers to interface with communication modules and/or socket of IC (Intelligent Card) cards. The other is use opto-couplers to isolate the metering chips and the MCU such that the MCU is not floating on the live wire but share the ground with communication modules and/or socket of IC cards. There are pros and cons of each method. For the latter method, the MCU and energy metering chip must be powered in different domain, as energy metering chip must connect directly to power line. The following figure shows the block diagrams of single-phase meter and poly-phase metering having different power domain. In this kind of meters, the MCU and metering chip have to be in different power domain, and opto-couplers are inserted between MCU and metering chip for communication (Figs. 1 and 2).

In the application that needs isolation between metering chip and MCU, every metering chip is powered independently. There are typically two kinds of power supply circuits. Power transformer (PT) based power supply and capacitor-divider network based power supply as shown in the following figures (Figs. 3 and 4).

K. Yang (✉) · S. Zhong · Q. Kong · C. Men
Vango Technologies, Inc., Hangzhou, China
e-mail: yangk@vangotech.com

S. Zhong
e-mail: zhongsp@vangotech.com

N. N. Tan
Institute of VLSI Design, Zhejiang University, Hangzhou, China

N. N. Tan et al. (eds.), *Ultra-Low Power Integrated Circuit Design,*
Analog Circuits and Signal Processing 85, DOI 10.1007/978-1-4419-9973-3_7,
© Springer Science+Business Media New York 2014

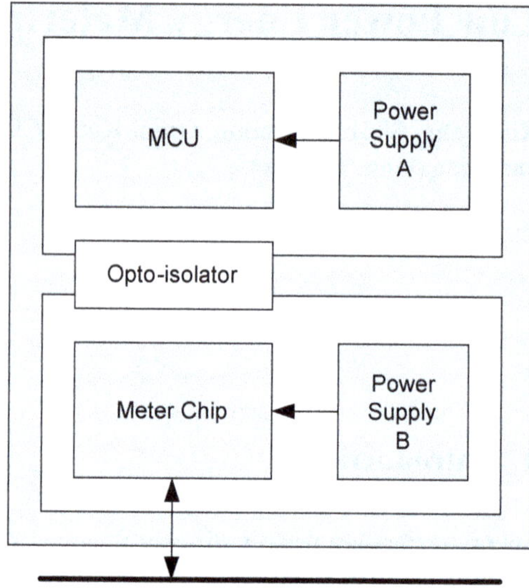

Fig. 1 Single-phase energy meter, with separate MCU and metering chip

Compared to the PT based power supply, the capacitor-divider network based power supply has weaker driving capacity but better reliability, such as electromagnetic compatibility performance. And more over, the capacitor-divider network based power supply is a more cost-efficient way.

In the capacitor-divider network based power supply, the capacitance of C6 is key to the driving capacity and reliability. The bigger the capacitance of C6, the more driving capability the power supply has but the worse reliability. Apparently using smaller capacitor offers better reliability and lower cost. But it calls for ultra-low power metering chips. Most low power energy metering chips consume a cur-

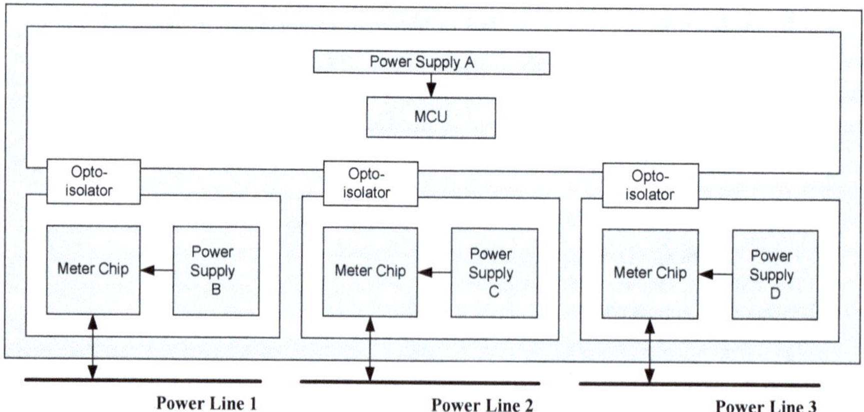

Fig. 2 Poly-phase energy meter, with separate MCU and metering chips

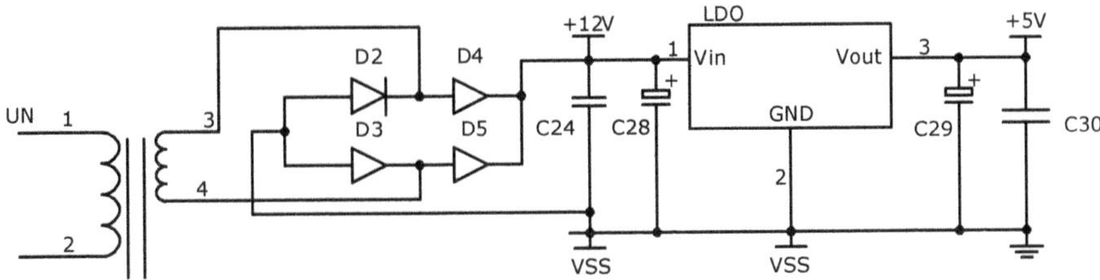

Fig. 3 Power transformer (PT) based power supply

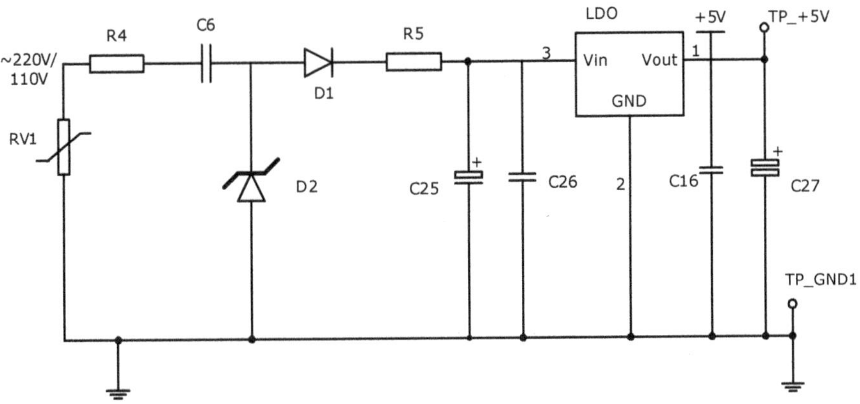

Fig. 4 Capacitor-divider network based power supply

rent north of 3 mA, too high for reliable capacitance-divider based power supply scheme. The current dissipation of the metering chip has to be at least halved.

In this chapter a low power single phase energy metering chip is introduced. It consists of three low power ADCs (Analog-Digital-Converter) to convert analog signals to digital domain, a energy metering application specific hardwired DSP (Digital Signal Processor) to process such signals, a system control module to manage power modes, and a UART (Universal Asynchronous Receiver/Transmitter) based peripheral for communication. The chip consumes only 1.5 mA at full-function meter mode, and 13 μA average power for current detection in anti-tempering mode. This chip is currently in production.

Low power consumption is realized in the following aspects:

- Low power metering modes design
- Low power metering algorithm design
- Low power DSP architecture design
- Low power circuits design
- Low power clock network design

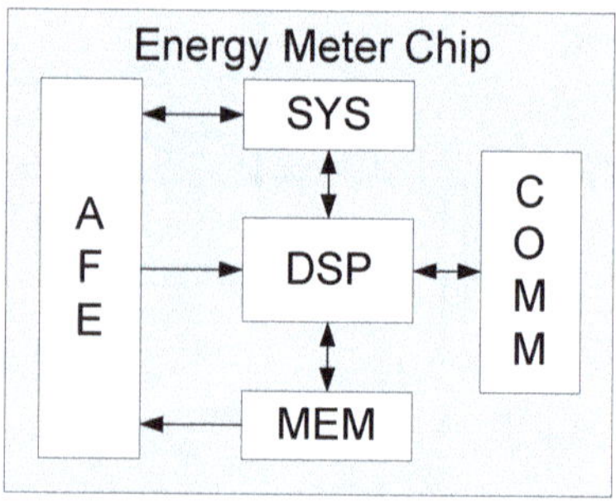

Fig. 5 Block diagram of the energy metering chip

2   Architecture of Energy Metering Chip

2.1   Overall Diagram

The overall diagram is shown in the following figure. The energy metering chip is divided into five major blocks.

- Analog unit (AFE)
- DSP unit (DSP)
- Communication unit (COMM)
- System management unit (SYS)
- Storage unit (MEM) (Fig. 5)

2.2   Analog Unit

The block diagram of the analog unit is show in the following figure (Fig. 6).

Three high precision sigma-delta ADCs are integrated into energy metering chip. One is used to convert the voltage signal, one is used to convert the current signal, and one is used to convert a multi-function signal such as temperature, internal/external DC signals. The ADCs can sample at 800 or 200 KHz, for higher precision or lower power consumption respectively. To further improve the noise performance, low noise amplifiers precede the ADCs.

There are three clock generators within the analog unit. One uses external crystal input to generate precise 3.2768 MHz clock. The other two clock generators use internal resistance and capacitors (RC) to generate 3.2 MHz/32 KHz clock signals. All the clock signals are managed by the system.

The analog unit also has a power management block. The ultra low power low drop linear regulator (LDO) generates 1.8 V power supply for digital circuits. It only dissipated 2-uA standby current. 3.3 V power input is used as supply for IO and analog unit. A POR (Power on Reset) and power down detection circuit is integrated in the power management unit. The band-gap circuit provides a 1.2 V reference voltage with a typical temperature drift of 10 ppm/°C. The temperature sensor makes it possible to calibrate the reference for a higher energy metering precision.

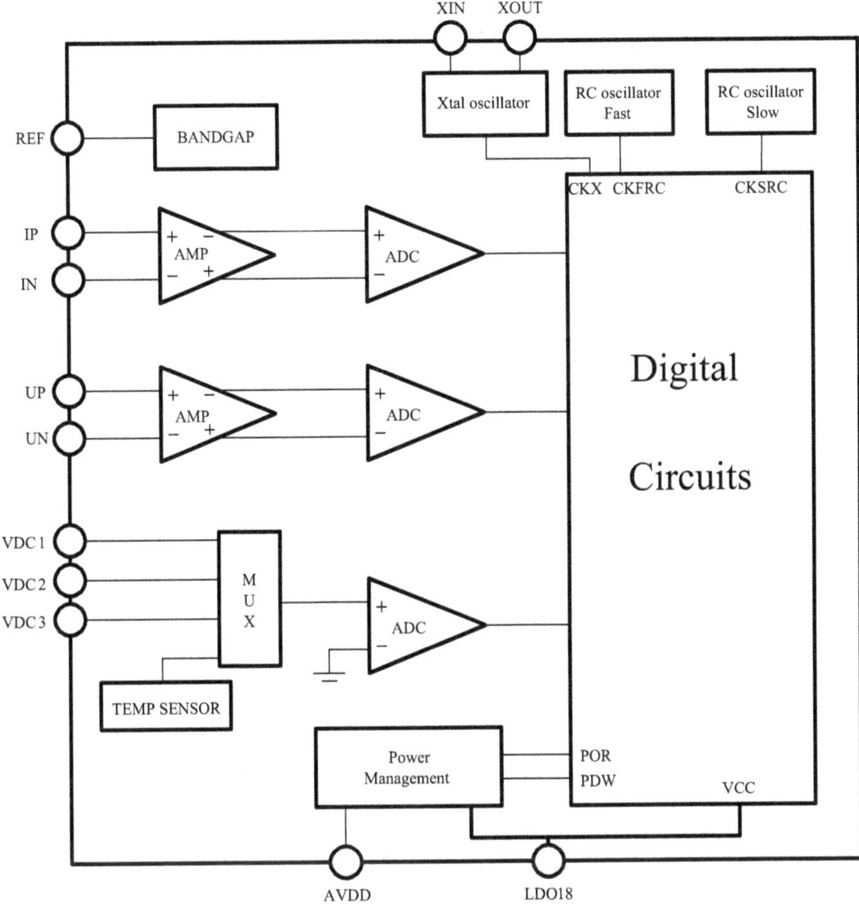

Fig. 6 Block diagram for analog unit

2.3 DSP Unit

The DSP is specifically design for power metering applications. Instruction set, register file, ALU (Arithmetic-Logic Unit) and bus architecture are all dedicatedly designed for power metering algorithm and for low power consumption. The DSP samples the data from ADC and provides power/energy, RMS (Root Mean Square) values, frequency of voltage signal (Fig. 7).

2.4 Communication Unit

Communication unit maintain a protocol based on a two-wire low speed UART transaction. Through this interface, the MCU on the energy meter can configure

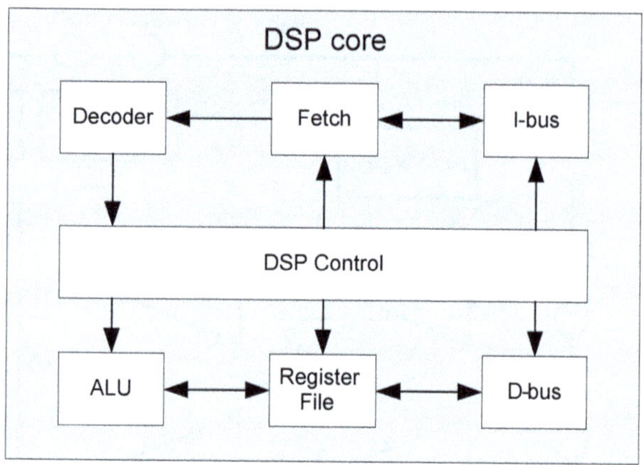

Fig. 7 Block diagram for DSP unit

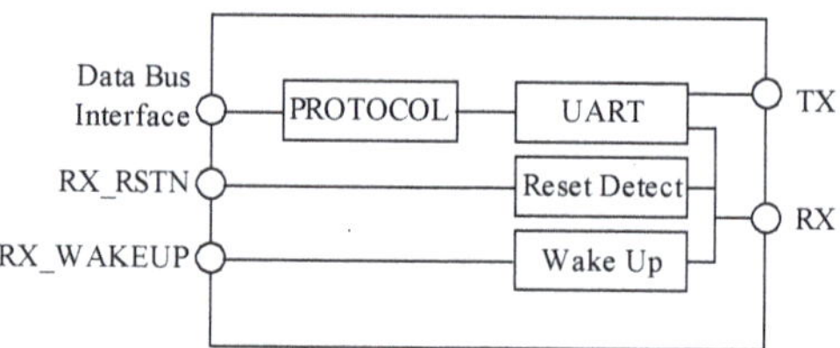

Fig. 8 Block diagram for communication unit

the chip and retrieve data from the chip. The RX-pin of the interface also serves as wake-up and reset pin, which can save MCU IO (Input Output) resource and reduce BOM (Bill of Material) cost (Fig. 8).

2.5 *System Management Unit*

System management unit maintains the function mode of the chip to achieve low power applications. It works on a 32 KHz clock and consumes very little power (Fig. 9).

2.6 *Storage Unit*

Storage unit consists of a SRAM (Static Random Access Memory) and some registers. The SRAM is the data memory to realize DSP algorithm and the register stores the configuration information for all other parts of the chip.

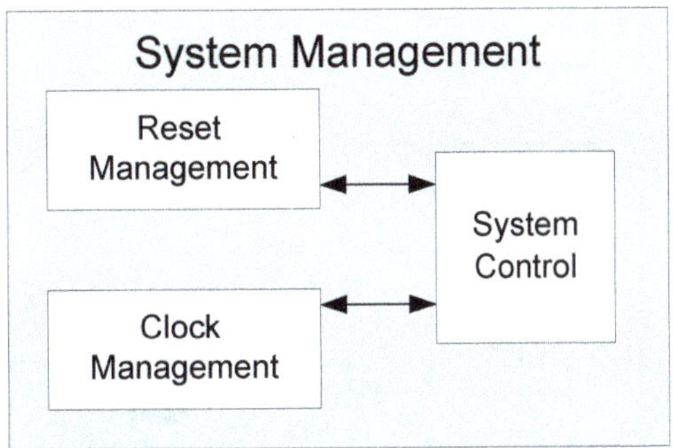

Fig. 9 Diagram of system management unit

3 Low Power Design and Consideration

Low power design is considered at the following levels:

- Low power metering modes design
- Low power metering algorithm design
- Low power DSP architecture design
- Low power clock network design
- Low power circuits design

3.1 Metering Modes Design

The energy metering chip samples signals of voltage and current, and provide following information:

- Signal waveform data, including original data, AC (Alternating Current) components and DC (Direct Current) component
- RMS value of the signal
- Active power and re-active power of voltage/current signals
- Energy of active/re-active power
- All above-mentioned information for fundamental (50/60 Hz) component of voltage/current signals
- Frequency of fundamental component of the voltage signal

The clock frequency and function of chip can be configured as different metering mode to realize different power consumption (Fig. 10).

Full Function, Full Precision Mode

In the full function/full precision mode, the chip works at full speed and provides all possible information at highest precision. In this mode, the chip consumes a maxi-

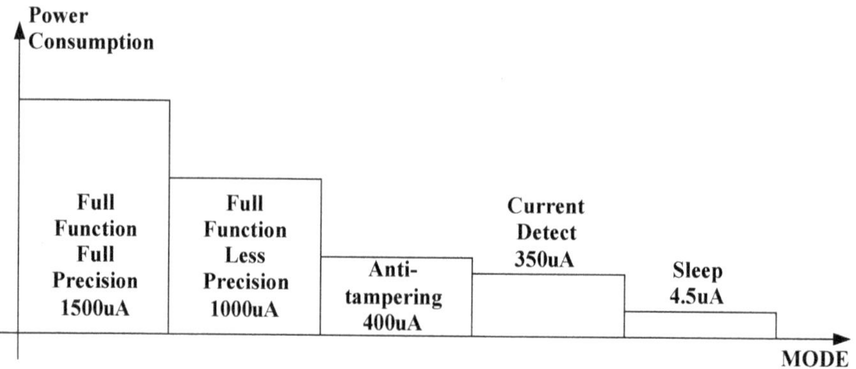

Fig. 10 Power consumption on difference mode

mal current of 1.5 mA. This mode is used when the energy meter is well powered and needs the highest metering precision.

Full Function, Less precision Mode

The full function/less precision mode is used in low power/low cost application where precision is not the highest priority. In this mode, the sampling rate and the working current of the ADC is cut down to save power consumption, while the DSP functions are all enabled to provide all possible information. In this mode about 1.3-mA current is consumed by the chip.

Less Function, Full Precision Mode

The less function/full precision mode is used in low power/low cost application where not all functions are necessary. In this mode, some of the DSP functions are disabled to save power, but basic energy metering information such as power/RMS are provided. In this mode about 1-mA current is consumed by the chip.

Anti-Tampering Mode

A typical tampering method is to cut the voltage input of the energy meter to make the chip see a voltage of 0 V. In this case, the chip still can see a current signal. Since power equals to voltage multiplied by current, the chip meters the power of zero. To avoid this, the chip needs to use RMS of the current signal as a virtual power and provide this power information. As voltage signal is cut to zero, the energy meter cannot get power from voltage signal. The energy meter can either get power supply

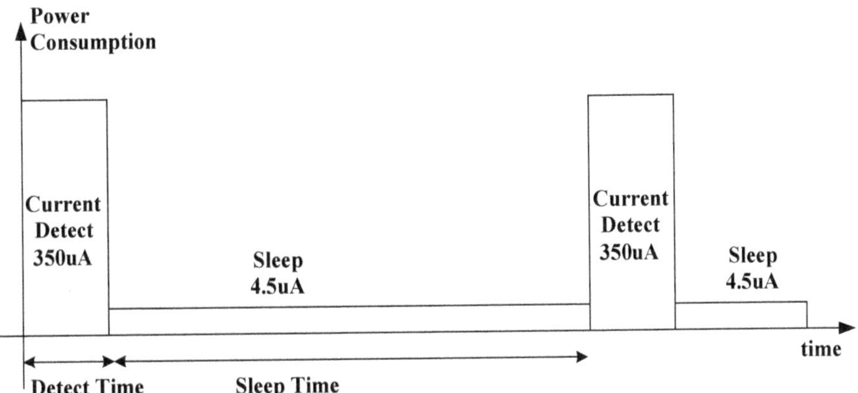

Fig. 11 Periodic current detection

from battery or from an inductor coil. To extend the battery life or to save the cost of the inductor coil, low power consumption is necessary in this case. In this case, the voltage ADC and the multi-function ADC are shut down; the sampling rate and the working current of the current ADC is cut down; the DSP function is reduces to only provide RMS of current signal; the power consumption of energy metering chip is reduced to about 400 μA.

Sleep Mode

When there is no voltage signal or current signal, the chip can go to sleep mode, in which all data are kept in SRAM and registers. All ADC is shut down and DSP is stopped but the LDO and lower frequency clock as well as the system monitoring circuits are still working. In this mode, the chip only consumes about 4.5 μA.

Current Detecting Mode

In the anti-tampering mode, the current signal can be 0. If the energy meter is working in full speed with the power supply from the battery, it will use up the battery in a couple of days. To avoid this, current detecting mode is introduced. Under current detecting mode, the chip is waken up periodically and works for a short period of time to monitor the original waveform of the fundamental signal. If the value of the waveform crosses the pre-set threshold, the chip will switch to anti-tampering mode, otherwise it will sleep again. In the current detecting mode, the chip only consumes around 13 ~ 850 μA, depending on how often the chip wakes up to monitor the current (Fig. 11).

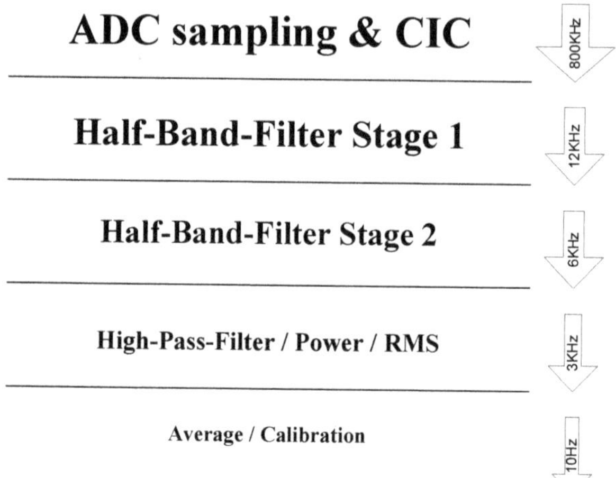

Fig. 12 Multi-rate processing scheme, the smaller font, the lower frequency

3.2 Metering Algorithm Design

A low power metering algorithm means the algorithm consumes as little horse power of DSP as possible. Here are some principles for low power algorithm design:

- Use multi-rate design [1]
- Use shift/add instead of multiply
- Tradeoff in filter design
- Use fix-point than floating-point design

Multi-Rate Processing

Sigma-delta ADC needs a CIC (Cascaded Integrator-Comb) filter to filter out high frequency noise. CIC filter is a typical multi-rate processing, which consists of an integrator and a differentiator. The integrator operates at ADC's sampling rate of 819.2 KHz. With a down-sampling factor of 64, the differentiator works on a much lower frequency of 12.8 KHz. Through the CIC filter, the data rate is reduced by 64 times.

Half-band-filter is another down-sampling filter that reduces the data rate by two.

In metering algorithms, there are a lot of processes to extract DC/AC part of signals through high-pass/low-pass filters. In the metering application, the components of signals are DC@0 Hz, fundamental@50 Hz (or 60 Hz), harmonic@100/150/200 Hz, etc. In this case, CIC filter is a ideal candidate to realize high-pass/low-pass filters, which makes the whole metering algorithm work in a multi-rate manner (Fig. 12).

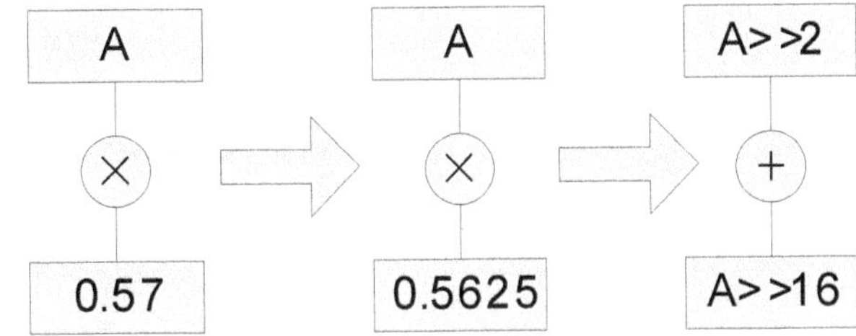

Fig. 13 Example for multiplication replacement

Multiplication Replacement

Multiplication is frequently used in filter design. But multiplication is much more complex than adder or shifter in circuit design and consumes more power. In a low power design, we must reduce the usage of multiplication and use adder/shifter instead. The following are two examples.

In different energy meter applications, we need to adjust the AC amplitude of input signals. In this case, we can use X2,/2, X4,/4, X8,/8 to realize a rough adjustment, which need only shift operation. And a fine adjustment can be realized in calibration, which need multiplication. As we discussed in the previous section, the calibration functions are at a very low frequency.

In metering algorithms, operation that in the form of "$Y = A \times X$" cannot be avoided. But we can carefully design the factor A to avoid usage of multiplication. For instance, if A is 0.57, we can adjust it to 0.5625, which is 9/16, "9/16X" equals to "1/2X + 1/16X" (Fig. 13).

Trade off in Filter Design

In low power energy metering applications, metering performance and power consumption must be balanced carefully. Extreme metering performance is never the only design goal. To meet low power requirement, we must use the simplest filter that can meet the metering requirement.

For example, a 90-degree phase needs to be inserted between voltage and current signals to calculate reactive power. There are two typical filters that can realize such requirement. Differentiator filter and Hilbert filter. The differentiator filter is very simple, but the amplitude-frequency response is not as flatten as Hilbert filter. But in current energy metering applications, the required precision of reactive power is much lower than the one of active power. So we can implement differentiator filter in the energy metering algorithm for lower power consumption.

Fix Point Implementation

The required dynamic range of current in a energy meter is typically 1000:1 with a accuracy of about 1‰. This leads to a minimal bit width of 21 bits. Considering

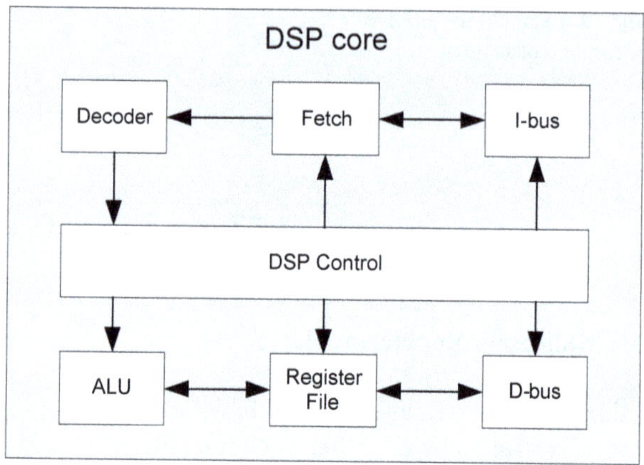

Fig. 14 Low power DSP architecture

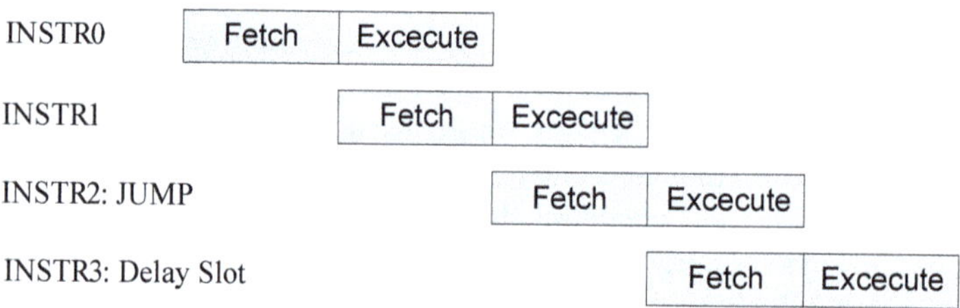

Fig. 15 2-stage pipeline

noise, input signal and internal feed-back filter implementation, a 32bit fix point algorithm is enough to meet the accuracy and dynamic range requirement. A lot of CIC filters are used in the algorithm, they are very fix-point-friendly filters.

Low Power DSP Architecture Design

A specifically optimized DSP architecture is designed for low power energy meter algorithm implementation. Its diagram is shown in the following figure (Fig. 14).

Every clock cycle, DSP fetch an instruction through the instruction bus (I-bus) interface and send it to the decoder module. The DSP controls the register file to realize data access through the data bus (D-bus) interface or ALU operations in ALU module.

The DSP works in a 2-stage pipeline, instruction fetch and instruction execute as shown in the following figure. If there is a branch instruction, a delay slot must be added after the branch instruction to be executed. A delay slot can be any instruction, if no instruction is needed to be executed, a NOP can be inserted (Fig. 15).

Fig. 16 Register file with
two 64-bit data register and
one 1-bit condition register

RegAH, 32Bit
RegAL, 32Bit
RegBH, 32Bit
RegBL, 32Bit
RegC, 1Bit

Fig. 17 Source and destination of ALU

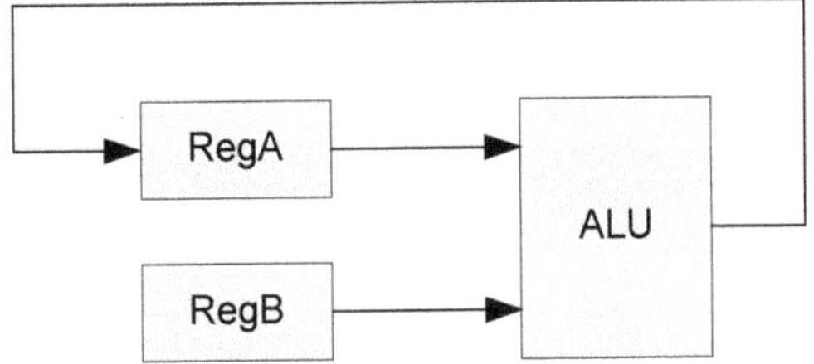

Register File

The register file is shown in the following figure (Fig. 16).

Only two 64-bit registers RegA/RegB and one 1-bit register RegC are integrated into the register file. The two 64-bit registers are used as the source and destination for ALU and can realize data load/store from and to the memory. Most operations in the algorithm are of 32 bits, but CIC filter sometimes need more than 32 bits. That's why we need 64-bit RegA and RegB. The 1-bit register is used as branch condition flag.

ALU

ALU unit processes data from RegA and RegB and send result back to RegA as shown in the following figure (Fig. 17).

Through profiling the metering algorithm, we design the ALU with following features to be the most suitable module.

The ALU can realize complex add/sub in 64-bit manner. Such an add/sub unit is combined with additional shift and anti-saturation options. As we discussed in the previous sections, to reduce power consumption, the metering algorithm majorly consists of add and shift operation. And to avoid saturation in the filters, anti-saturation operation is necessary for output and every feedback nodes of the filter. A complex add/sub unit can realize those three operation in a single clock as shown in the following figure. This reduces the clock frequency of the DSP and the number of register operation (Fig. 18).

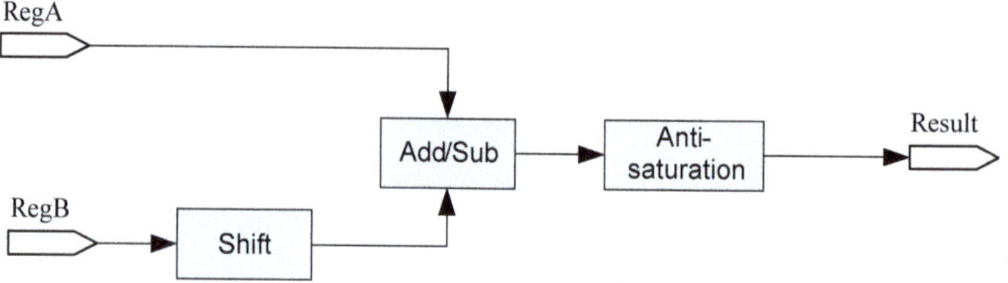

Fig. 18 Complex add/sub

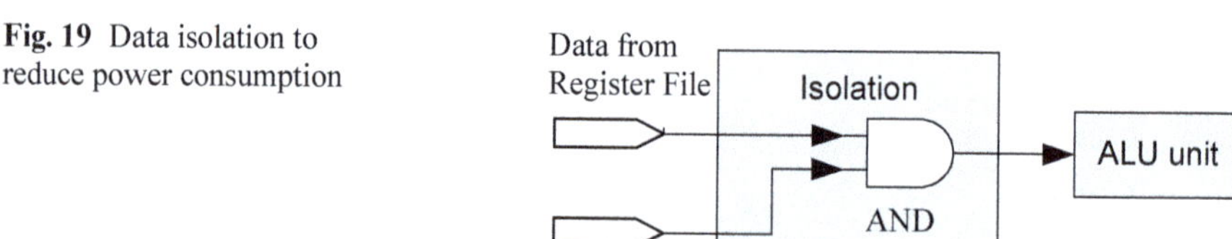

Fig. 19 Data isolation to reduce power consumption

The ALU can realize multiplication and square root in a slower manner. At 3,200 Hz rate-level, only several multiplications operations are needed for power and RMS calculation. Other multiplication operations for calibration and so on are executed at 50 or 12.5 Hz rate-level. In the case of square root, the operation is only at 50 Hz. To reduce power consumptions, the multiplication unit is designed as a 32×32bit multiplication generating a 64-bit result in 32 clock cycles. To balance the area and accuracy, the square root unit is designed as a 50-bit square root generating a 25-bit result in 26 clock cycles.

To avoid unnecessary power consumption in ALU, every major arithmetic unit is integrated with an isolation unit to tie the input to zero when there is no corresponding operation to eliminate unnecessary circuit switches [2]. This is illustrated in the following figure (Fig. 19).

Bus Architecture

The bus of DSP is of Harvard architecture with a 16-bit instruction bus and a 32-bit data bus. The bus architecture is shown in the following figure (Fig. 20).

Most operations in the metering algorithm are if 32 bits. The 32-bit data bus can achieve higher power efficiency than 64-bit data bus. And we can realize a 64-bit bus access through two 32-bit load/store operation.

Of the instruction bus, there is only one ROM (Read Only Memory) stored instruction for the DSP.

On the data bus, there is a SRAM, a set of register pool and some peripherals. SRAM acts as data for metering algorithms. Register pool stores configuration for

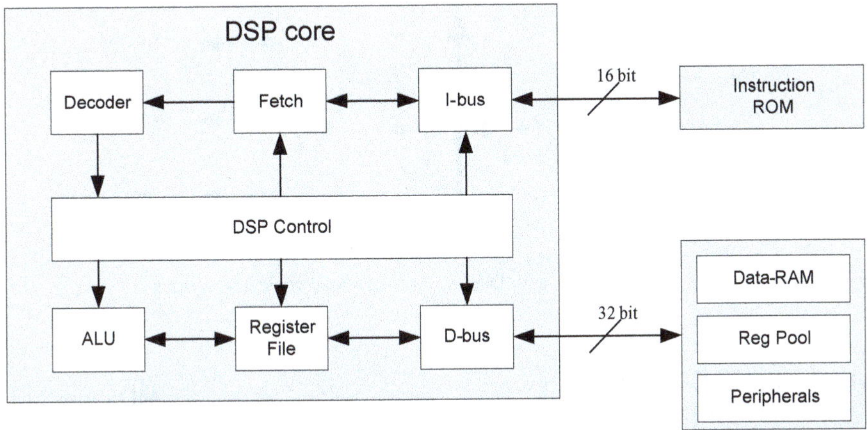

Fig. 20 Bus architecture of metering chip

DSP, analog front end and communication. Peripherals include some external modules for metering application.

Instruction Set Architecture

The instruction set is a highly optimized 16-bit RISC (Reduced Instruction Set Computing) Instruction set architecture (ISA).

NOP

The NOP (No Operation) instruction does nothing. It is used to fill the space for multi-cycle operations such as multiplication or to fill the delay slot in the branching operation.

Memory Access

LOAD/STORE instructions are used to exchange data between the registers and the memory. As there are only two 64-bit registers in the register file, only direct addressing mode is supported.

A SWAP instruction is used to swap information between regA and regB. As all ALU instruction is in the mode of "regA=ALU(regA, regB)". A SWAP operation can reduce the need for load/store operation (Fig. 21).

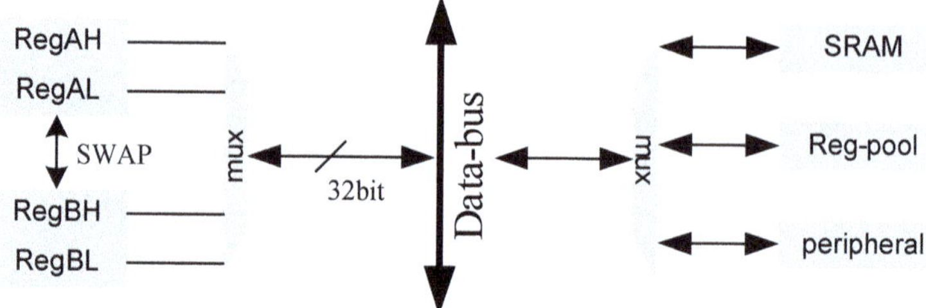

Fig. 21 Register file exchange data with Data bus through load/store instructions

Fig. 22 Multi-cycle multiplication instructions

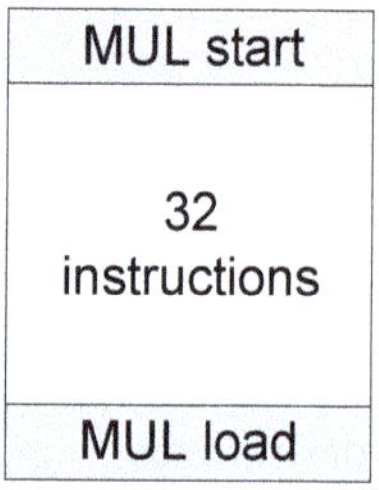

ALU

ARC (Arithmetic Command) realizes a one-clock-cycle arithmetic operation for the complex add/sub operation. It is a combination of add/sub, shift and anti-saturation.

Multiplication/square root (MUL/SQT) realize multi-clock-cycle arithmetic operation for multiply and square root. Such operation starts with a START option and, after a curtain number of clock cycles, ends with a LOAD operation to load the result to the register. Between START and LOAD, any other instructions can be inserted to realize a parallel processing. If there is no other instruction to be executed, NOP must be inserted instead (Fig. 22).

Branching

There are unconditional jump instruction JMP and register RegC based conditional jump instruction JPC in the instruction set. There are two stages in the DSP pipeline, when executing the branch instruction, the next instruction is already fetched into the decoder, so a delay slot must be inserted after every branch instruction. The delay can be replaced by any instruction, if there is nothing to do, a NOP can be used as delay slot instruction. This is shown in the following figure (Fig. 23).

A special conditional jump flag instruction CND is introduced in this instruction set. CND is used to compare an internal counter with a given condition. This instruction is specifically designed for multi-rate realization. Through this we can easily design real-time multi-rate program (Fig. 24).

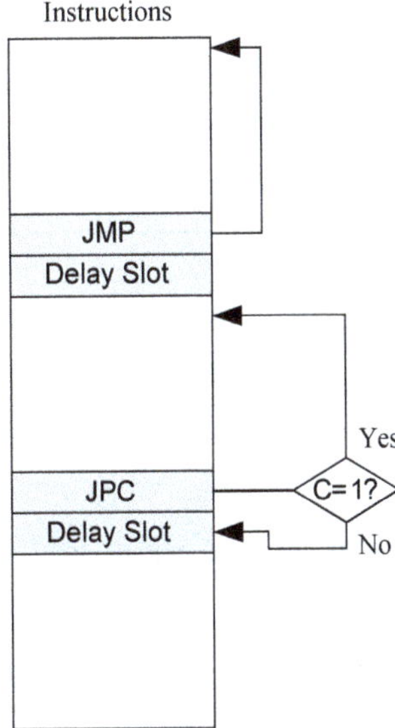

Fig. 23 Unconditional and conditional branch, with delay slot inserted

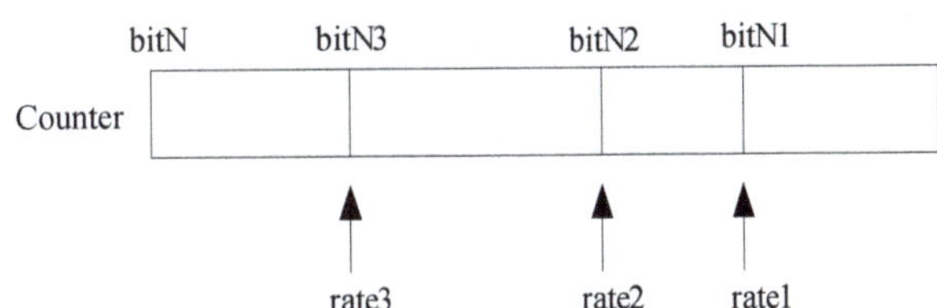

Fig. 24 CND for multi-rate implementation

Reliability Consideration

Like most computer architecture, this energy metering DSP works basing on a Program Counter (PC), which increments by 1 for no-branching instruction and set to a destination address for branching instruction every clock cycle. For reliability consideration, typical MCU/DSP system introduces a Watch Dog Timer (WDT) unit. Through resetting the WDT, the MCU/DSP realizes a "Feed Dog" operation. If the WDT is not fed on time, it will reset the whole MCU/DSP system and reset PC to zero. The necessary feeding frequency is about 1 s. To enhance the reliability, this DSP introduces a similar method, but there is no need for "Feed Dog" operation. The WDT of this DSP resets the PC every 256 clock cycles. This makes the maxim number of fault operation is 255 and makes it a more stable design.

3.3 Low Power Communication Unit

MCU and metering chip are isolated, the communication between them need opto-coupler for each signal as shown in the following figure. The opto-coupler circuit needs a pull-up resister or pull-down resister that will consume extra power in some

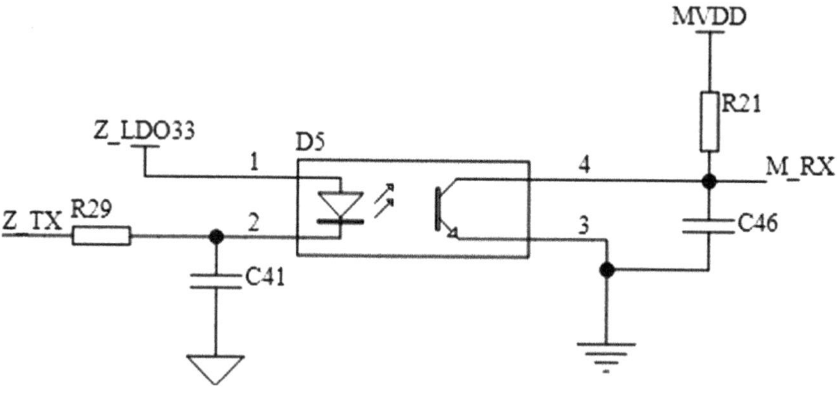

Fig. 25 Opto-coupler circuit

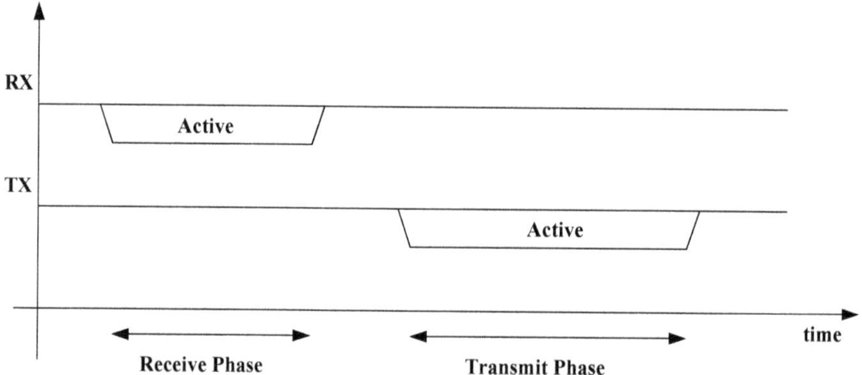

Fig. 26 Half-duplex communication, only one of RX/TX is active at one time

case. The key to save communication power is to reduce the number of opto-coupler (Fig. 25).

A half-duplex UART communication is used in the metering chip. Only 2 opto-couplers are need for this communication protocol. And as the data transaction is in a half-duplex way, only one opto-coupler may consume energy at one time (Fig. 26).

3.4 Low Power Analog Circuits Design

Most analog circuits have the same architecture as the one in the energy meter SoC in other chapter. In this energy metering chip, the most power consuming part, analog-digital-convert is optimized for lower power consumption. The following table shows current dissipation for each part of analog circuits (Table 1).

Table 1 Power consumption of analog circuit

Unit	Operation current (μA)
Crystal oscillator (6.4 MHz)	~120
Global bias	~70
Band-Gap	~80
High-speed RC oscillator	~80
Voltage ADC	~300
Current ADC	~400
Multi-function ADC	~450

In most cases, the multi-function ADC is power off. Only in those applications that need extreme precision, the multi-function ADC is power on to measure the temperature and to calibrate the band-gap according to the temperature.

To further save power, the bias current of ADC can be reduced into two levels, the SNR of ADC will become worse, or in another words, we have to sacrifice metering performance for lower power consumption.

For reliable UART communication, we will need a stable clock. In case there is no external crystal oscillator, an internal clock has to be generated. Such internal RC oscillator is shown in the following figure (Fig. 27).

The high-speed RC oscillator is designed to generate a backup 3.2 MHz clock for external 3.2 MHz crystal. For a reliable UART communication, the maxim baudrate error is 5 %. The RC oscillator is integrated with a trimming feature to be configured to 3.2 MHz $\pm$ 2 %, and can maintain a frequency of 3.2 MHz $\pm$ 3 % from -40 to 85 °C. This RC oscillator consumes less power than external crystal oscillator, and makes the chip more reliable in case the external crystal is damaged.

3.5 Low Power Clock Network Design

An un-optimized clock network can consume 30 to 50 % power of the whole chip [3, 4]. The situation goes even worse for those slow modules like UART, SPI (Serial Peripheral Interface), RTC (Real Time Counter), etc. Clock gating is the most efficient methodology to reduce the power consumption of clock network. In the metering chip design, four stages of clock gating are introduced.

Clock Gating for Conditional Registers

By default, a conditional register can be implemented with a mux (multiplex) in front of a standard flop, as shown in the following figure. The register updates data from D when condition signal "cond" is true and keeps its value otherwise. For those conditional registers that only need to be updated occasionally, most of the power occurs in the associate clock network. Use clock-gating unit to realize the conditional function is a good alternative for further power saving (Fig. 28).

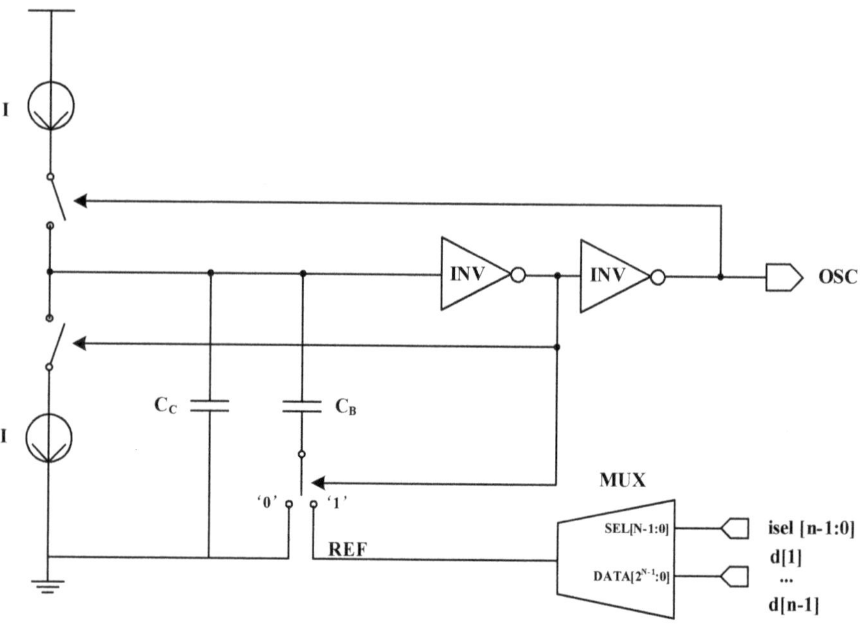

Fig. 27 High-speed RC oscillator with trimming circuit

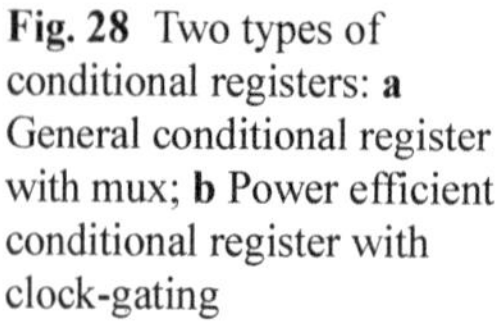

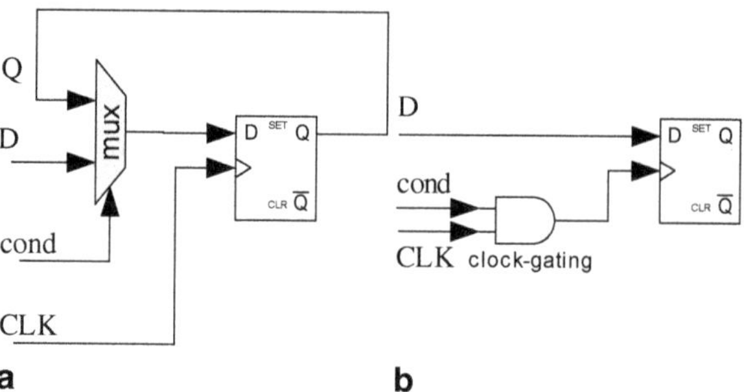

Fig. 28 Two types of conditional registers: **a** General conditional register with mux; **b** Power efficient conditional register with clock-gating

The clock-gating part of conditional register is not just a single AND, it must be a special design unit to avoid glitch on clock network. The following is a latch based clock-gating cell. The latch updates data from EN when CLK is low and gate the CLK with an AND cell (Fig. 29).

Clock Gating for Modules

There are a lot of conditional registers in the modules. Every group of conditional registers has a clock-gating unit. These clock gating units are loads of the clock tree, and will keep switching even the according conditional registers are clock-gated. To

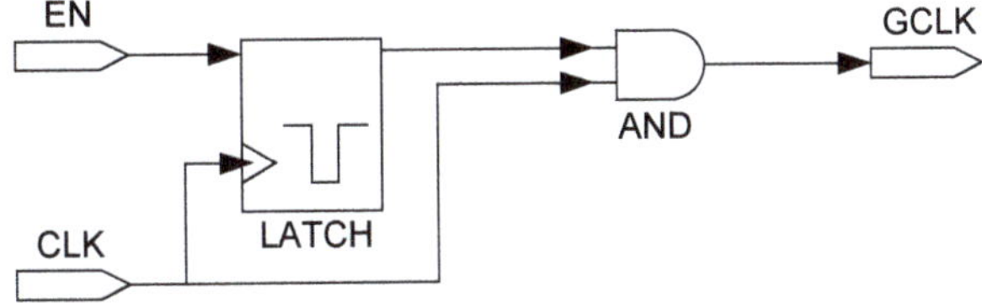

Fig. 29 Latch based clock-gating unit

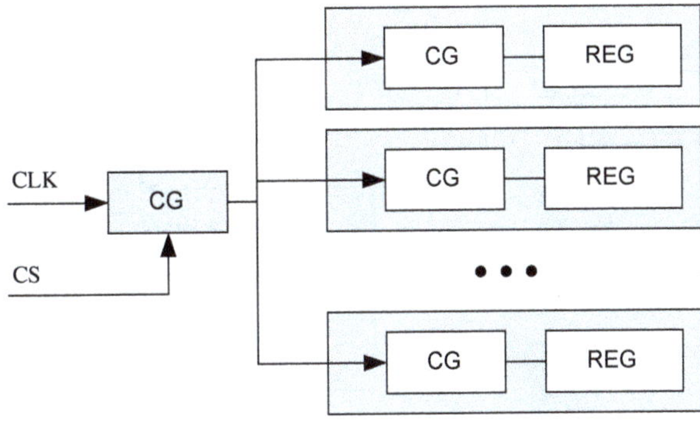

Fig. 30 Module level clock-gating. (*CG* clock gating module, *REG* register, *CS* chip select)

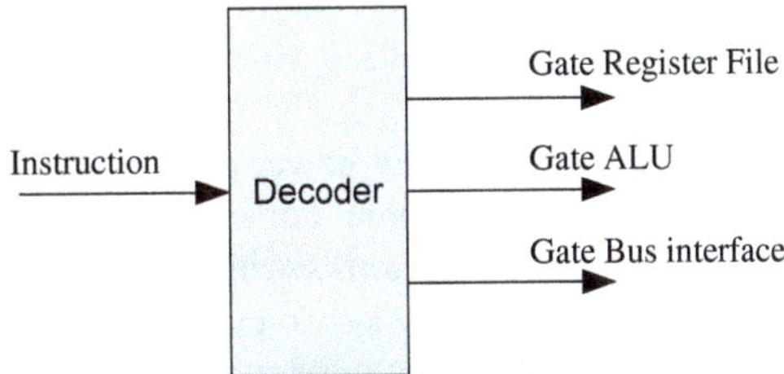

Fig. 31 Instruction level clock gating

reduce this part of power consumption, a module level clock gating unit is implemented to further save the clock network power consumption when the module is in idle status or unselected (CS=0). This is shown in the following figure (Fig. 30).

Clock Gating for Instructions

Not all modules need to be active for an instruction in the DSP. The decoder of DSP will find the necessary modules for current instruction and disable all unnecessary modules through clock gating. For instance, when executing the instruction NOP, no module is active, so all modules including ALU, register files, bus interface, etc. are clock-gated to save power. When executing a memory access instruction, ALU will be shut down through clock-gating to reduce power consumption (Fig. 31).

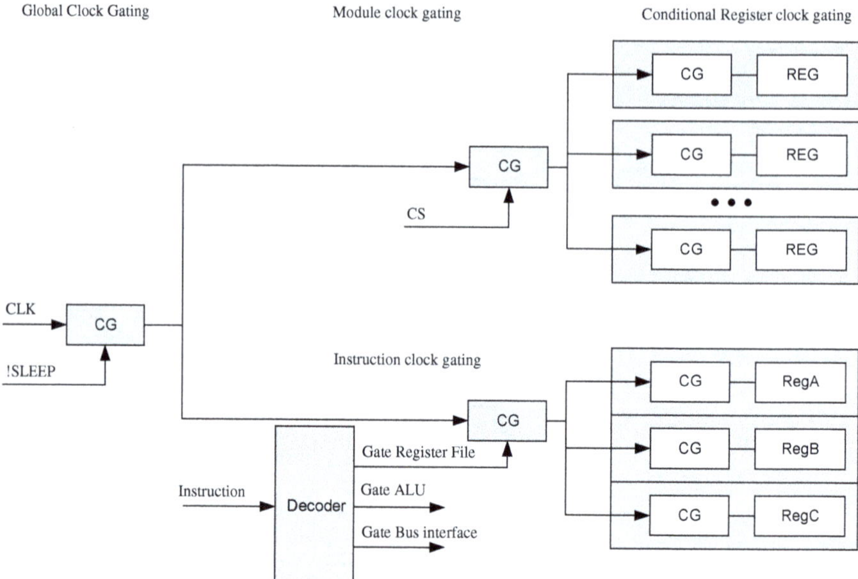

Fig. 32 Multi-stage clock gating scheme

Clock Gating for Working Mode

For some extremely low power oriented applications, the chip needs to consume as little power as possible. For instance, in sleep mode as we described in previous sections, the chip works with power supplied by a battery, in this case, ultra low power design is the key to extend the battery life. A global clock gating unit is implemented into the chip for such circumstance to cut down the whole clock network, and leave only leakage current in the chip.

Multi-Stage Clock Gating Scheme

The clock gating is designed to be 3-stages in the metering chip as shown in the following figure (Fig. 32).

The first stage is a global clock gating. In sleep mode, the clock network is gated to achieve ultra low power consumption at about 4.5 μA.

The second stage is module level clock gating. It is controlled through the module select signal for those peripherals such as the register pool, or is controlled by the decoding result of current instruction.

The last stage is the conditional register clock gating. It controls the clock network for every group of registers with the same condition.

Most registers in the metering chip update their data in a very slow frequency, only RegA and RegB in the register file of the DSP core update in a relatively

Fig. 33 Photo of metering chip

fast speed. Through the 3-stage clock gating, the power consumption of the clock network can be significantly reduced to a very low level less than 10 % power consumption of the whole chip.

4   Results

The metering chip is fabricated in TSMC mixed-mode technology and is assembled in a 16-pin SOP package. In a dynamic input range of 1:5000 of current signal, the chip's metering error is less than 0.1 %. This performance can be maintained during −40 to 85 °C working conditions. EMI rejection features can be proved to have 15 KV contact and non-contact discharge tolerance. It is the world's first single phase energy metering chip consumes only 1.5 mA with full function. The die photo is shown in the following figure. This chip is currently in full production (Figs. 33 and 34).

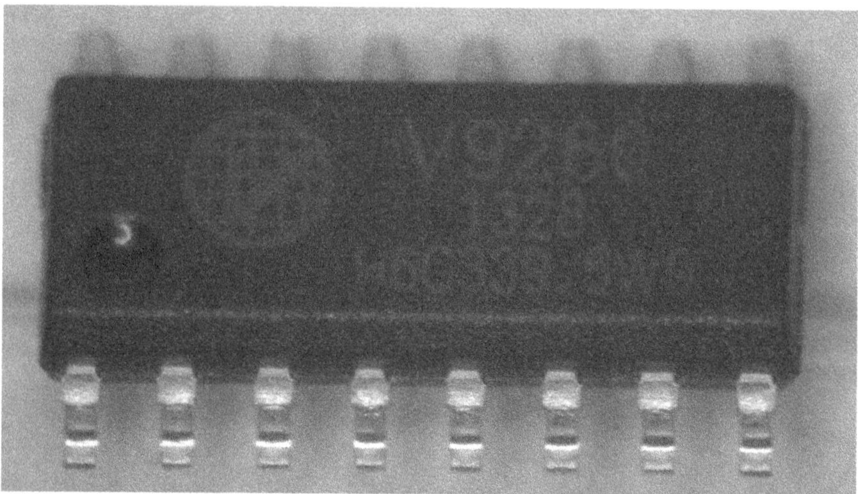

Fig. 34 Metering chip assembled in SOP16 package

References

1. Chih-hsien Kung; Devaney, M. J. "Multirate Digital Power Metering," Proc. of the Instrumentation and Measurement Technology Conference, 1995.
2. Chandrakasan A P, Srivastava M B, Brodersen R W. Energy efficient programmable computation [A]. In: Proceedings of International Conference on VLSI Design, Calcutta, India, 1994, pp 261–264.
3. J. Oh, M. Pedram. Gated clock routing for low power microprocessor design. IEEE Trans. On CAD, Jun-2001, Vol. 20, No. 6, pp 715–722.
4. Mohanty S P, Ranganathan N, Krishna V. Datapath Scheduling using Dynamic Frequency Clocking[A]. In: IEEE Computer Society Annual Symposium on VLSI, Pittsburgh, Pennsylvania, 2002, pp 58–63.

A Single-Phase Energy Metering SoC with IAS-DSP and Ultra Low Power Metering Mode

Yan Zhao, Shupeng Zhong, Kun Yang, Changyou Men, Nianxiong Nick Tan and Xueping Jiang

1 Introduction to Energy Metering ICs

Smart grid has become a revolution to our lives. Energy meters provide the basic information of smart grid, including active power/reactive power/apparent power and corresponding energy, RMS values of voltage and current, harmonic information, etc. The meter computes all these items and communicates with other meters or the gateway [1, 2]. Therefore, the integrated circuit chips inside the meter must have the ability to accomplish all these functions.

Traditional integrated circuit solutions for energy meters were composed of separate chips, usually a metering front end or an ADC plus an MCU [3–5]. An ADC simply converts analog signals such as voltage and current into digital bits. A metering front end can compute power or energy utilizing some filtering operation besides converting. Tasks for the MCU depend on the configuration of the converting parts. If a metering front end is used, the MCU just reads out the power or energy results from the chip via some serial communication port, and then does some interface and communication jobs. If an ADC is used, the MCU must accomplish by itself all the filtering operations and calculate all the power and energy results. In addition, there are still interface and communication jobs for the MCU to deal with.

A few years ago, energy metering SoCs were accepted by the utility industry. The SoC provides a single chip solution for energy meters. All ADCs, CPU core, metering algorithm, LCD driver, and communication ports are integrated into one single chip. The benefits compared with previous separate chip solutions are lower cost, higher reliability and better efficiency.

Y. Zhao (✉) · S. Zhong · K. Yang · C. Men
Vango Technologies, Inc., Hangzhou, China
e-mail: zhaoy@vangotech.com

N. N. Tan
Institute of VLSI Design, Zhejiang University, Hangzhou, China

X. Jiang
Smart Grid Research Institute, State Grid Corporation of China, Beijing, China

N. N. Tan et al. (eds.), *Ultra-Low Power Integrated Circuit Design,*
Analog Circuits and Signal Processing 85, DOI 10.1007/978-1-4419-9973-3_8,
© Springer Science+Business Media New York 2014

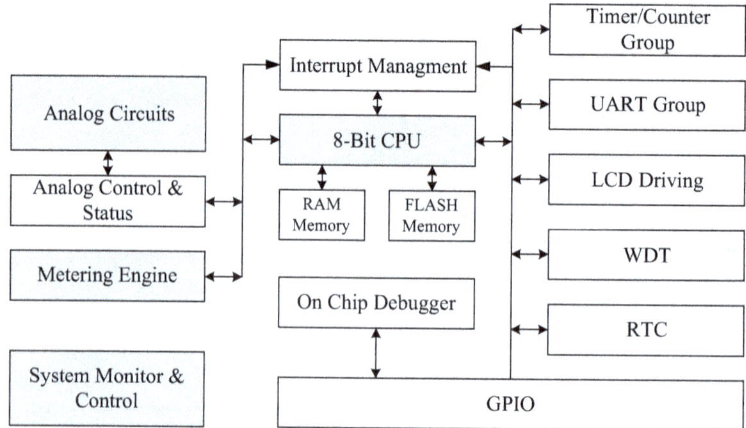

Fig. 1 Architectural diagram of the SoC

In most of energy metering SoC products, the metering engine is implemented by a universal embedded DSP processor. This design method has the advantage of being easy to upgrade and shorter time to market because all algorithms are firmware. But larger circuit area and larger power consumption are also inevitable because of redundancy and low efficiency.

In this chapter, an energy metering SoC with specifically designed low power metering engine and low power metering modes will be introduced [6]. The architecture and the low power design method of this SoC, especially of the metering engine, have been adopted by a series of energy metering ICs for both single-phase and poly-phase applications. Tens of millions chips have been shipped.

2 Architecture of the SoC

As an SoC, analog circuits, digital circuits, volatile and non-volatile memories are integrated. Figure 1 shows the overall architectural diagram of the SoC. The shadowed blocks are analog circuits, CPU core, metering engine, and system control respectively. Other blocks are peripherals of the CPU core and on-chip memories.

The 8-bit embedded CPU core is an Intel 8052 compatible core. This core is selected for its high code density, simplicity in hardware, and ease of use for almost all meter designers. As mentioned in Sect. 1, no complicated tasks are to be processed by the CPU core because a metering engine and various on-chip peripherals are integrated.

On-chip peripherals include a Real Time Clock (RTC), a Watch Dog Timer (WDT), a LCD driver, UARTs, Times/Counters, Interrupt Management, and an On-chip Debugger. All of them together help the SoC accomplish interface and communication jobs. A 4 K-byte SRAM memory and a 32 K-byte FLASH memory with

ISP (In System Programmable) and IAP (In Application Programmable) capability are also integrated.

Four channels of high precision sigma-delta ADC are used to convert one voltage input, two current inputs, and one versatile input such as battery voltage or temperature into digital bits. The ADCs have 20-bit in-band resolution in all temperature ranges (−40 to 85 °C). The oscillation circuit and PLL circuit provide 32,768, 204,800, 819,200 and 3,276,800 Hz clocks in low jitter to various blocks on the chip. Regulators give 2.5 V low-dropout output to digital circuits and 3.3 V low-drop out to IO (Input and Output) circuits. Temperature sensor enables the SoC to calibrate its RTC to counter the frequency drift of the external crystal's oscillation.

The metering algorithm is realized by the metering engine. In this SoC, the metering engine is implemented by a dedicated 32-bit fixed-point instruction and architecture specific DSP processor. The DSP processor is designed and optimized for multi-rate filtering operations and intensive computational operations. Section 4 will introduce the details of the DSP processor.

The system control block implements the SoC's low power strategy. According to the power supply source and the RMS values of voltage input and current input, the software running on the CPU core determines whether to go into some low power mode or not. The system control block gates some other circuit block's clock or control the flip rate of D-flip-flops in that block to realize low power demands. Section 5 will introduce how the low power strategy works.

3 Analog Circuits

The block diagram of analog circuits is shown in Fig. 2.

Analog circuits in this SoC are mainly composed of three modules; Power Management, Analog-to-Digital Converters, and Clock Management.

The Power Management module provides power supplies for both analog and digital circuits. The SoC chip is powered by a single 5 V voltage source, so Low Drop-Out linear regulators (LDO) are necessary. There are two LDOs in this module, providing separate voltages for analog and core digital. A Power-On Reset circuit and Power-Down Detection circuit are also integrated in this module.

The ADC module is composed of four high precision Sigma-Delta modulators preceded by low noise amplifiers; two of them quantizing two channels of current signals, one for quantizing the voltage signal, and the last one for versatile inputs such as temperature or battery measurement. There is a bandgap reference with 10 ppm/°C drift provide reference voltages for these four ADCs.

The clock management module is composed of a crystal oscillator, a Phase Locked Loop (PLL) and a Resistor-Capacitor (RC) oscillator. The crystal oscillator generates a precision clock with a frequency of 32,768 Hz. This clock is connected to PLL as the reference clock source. The PLL generates clocks for both digital and analog circuits. The output frequency of PLL could be selected by the CPU.

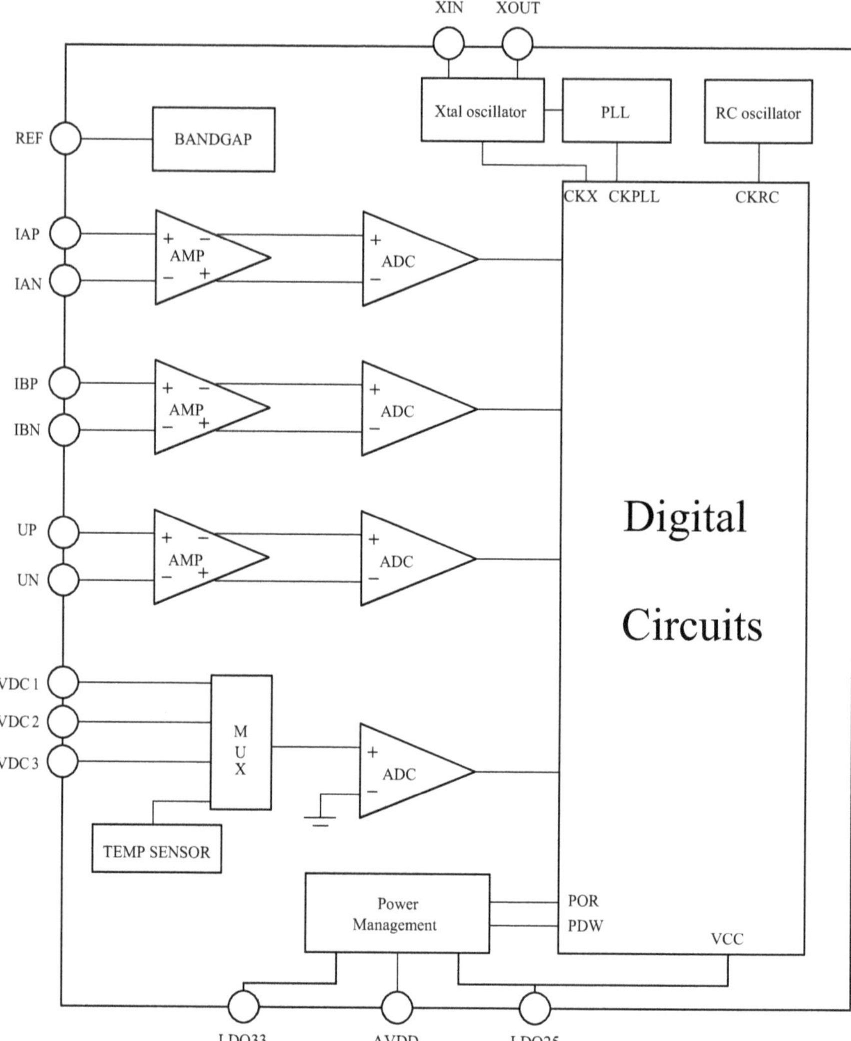

Fig. 2 Block diagram of the analog circuits

3.1 *Ultra Low Power LDO*

The LDO for the digital circuits consumes micro-ampere standby current. This ultra-low-power LDO is mainly used by the digital core circuits. The output voltage can be trimmed by the CPU to reduce power consumption of digital circuits. Because some of digital circuits should be always powered, the LDO should not be powered down even when the chip is supplied by battery. Ultra-low standby power is a key parameter. The LDO itself is designed as an ultra low power circuit with a standby current consumption of 1 μA.

Fig. 3 Ultra low power LDO

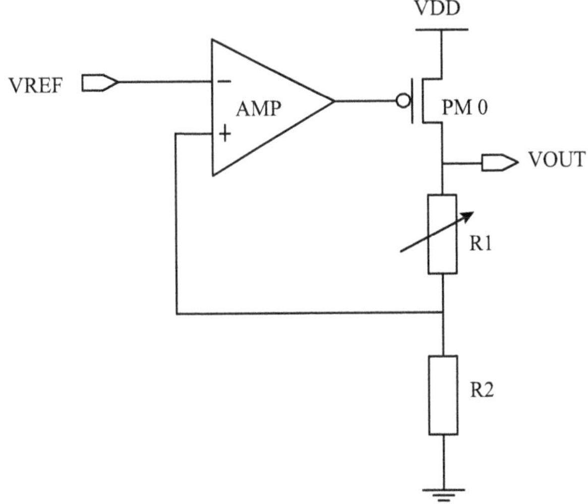

Fig. 4 Ultra low power
reference

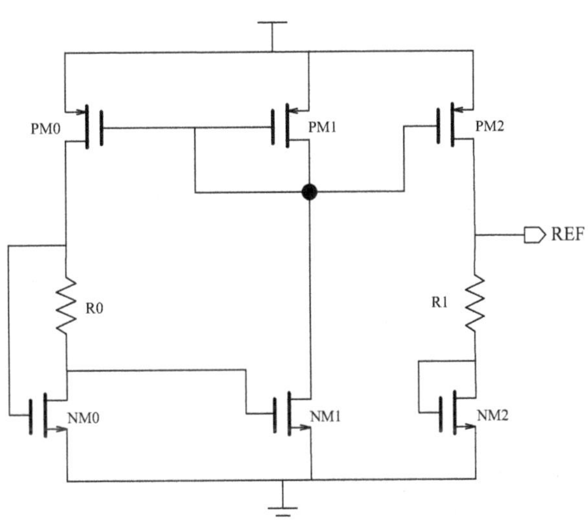

The structure of the LDO is shown in Fig. 3. This module is composed of error amplifier (AMP), driving MOS (PM0), and feedback resistors. The resistor R1 is adjustable by the CPU.

The input reference VREF is provided by an ultra low power reference. Unlike traditional bandgap references, the positive temperature coefficient of this low power reference is generated by two NMOS biased in weak inversion region. The block diagram of the ultra low power reference is shown in Fig. 4.

All the MOSFETs of Fig. 4 are biased in weak inversion region for ultra lower power consumption. The I–V curve of a weak inversion MOSFET is similar to that of a bipolar transistor, so the principle is also quite similar to a bandgap reference.

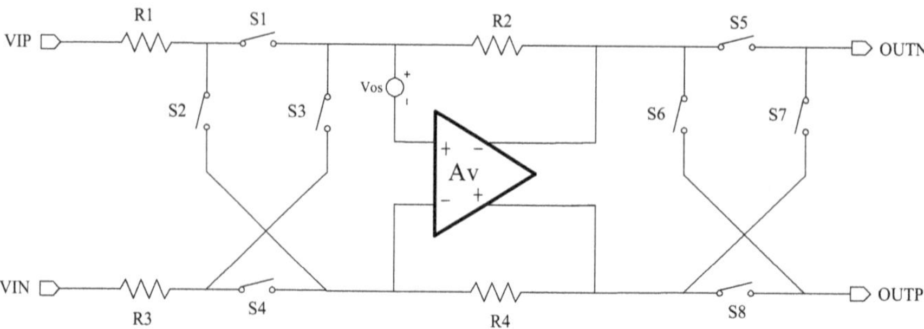

Fig. 5 Chopper stabilization amplifier

The positive temperature coefficient voltage is generated by $V_{GS,\,NM0} - V_{GS,\,NM1}$, and the negative temperature coefficient voltage comes from $V_{GS,\,NM2}$.

The current consumption of this reference is less than 1 μA, much lower than the traditional bandgap reference.

3.2 ADCs

ADCs in this SoC are one of the most important modules which determine the precision of metering. Each ADC is composed of two sub-modules, a low noise amplifier and a second-order single-bit sigma-delta modulator.

Low Noise Amplifier

The signal coming from current sensors are sometimes too weak to be directly quantized by analog-to-digital converters (ADCs) due to the inherent noise of any ADC, including KT/C noise, flicker noise, and thermal noise. Moreover, the signal frequency is located at 50 or 60 Hz, where large amount of noise exists. Thus, a low noise amplifier is necessary prior to the ADC.

The type of low noise amplifier used here is a chopper stabilization amplifier. The amplifier moves the signal away from its original frequency to a higher frquency by modulation, and demodulates them to move back after amplification. So the principle of a chopper stabilization amplifier is to spectrally separate the signal of interest and unwanted noise. The low-frequency noise that is modulated to higher frequency can be filtered out by a following low pass filter.

The block diagram of this amplifier is shown in Fig. 5.

It's composed of eight switches, four resistors and one operational amplifier. The four switches before the amplifier play a role of modulating the signal. The chopping clock has two phases named Φ1 and Φ2; switches S1/S4 are controlled by phase Φ1 and switches S2/S3 are controlled by Φ2. The other four switches

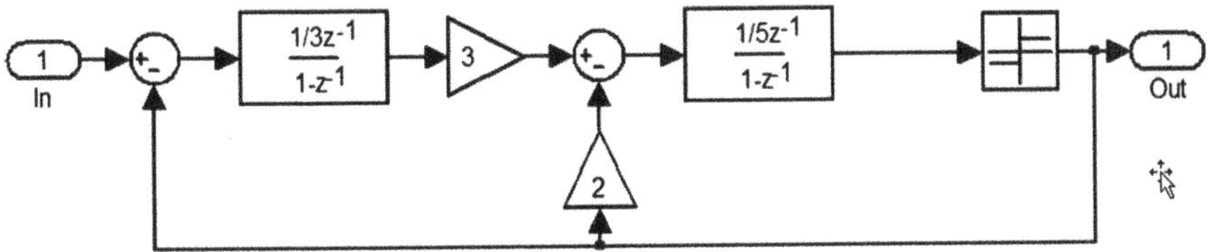

Fig. 6 Architecture of the sigma-delta modulator

act as a demodulator; switches S5/S8 are controlled by $\Phi1$ and switches S6/S7 are controlled by $\Phi2$. Resistors together with the operational amplifier amplify the modulated signal by a factor of $\dfrac{R2}{R1}$. (R1 = R3; R2 = R4)

Sigma-Delta Modulator

Sigma-delta modulators are widely used in high precision narrow band applications. In energy metering, the signal bandwidth is considerably low, only 1.6 KHz for the 31st harmonic. The sampling clock rate is 819.2 KHz in this chip, 256 times higher than the signal bandwidth, so a second order single bit sigma-delta modulator is suitable for this application.

The structure of this sigma-delta modulator in discrete time domain is shown in Fig. 6.

The modulator is composed of two integrators and one quantizer. The coefficients of these two integrators are not ideally one but 1/3 and 1/5, or else the output of these two integrators will saturate and drive the transistors in the integrators into deep linear status. This would severely decrease the performance of the integrators. The feedback factors for these two integrators are 1 and 2. So the expression of the noise transfer function is given by

$$\text{NTF} = (1 - z^{-1})^2 \tag{1}$$

for *NTF* expressing Noise Transfer Function, and z^{-1} expressing unit delay. It's a second-order high pass filter. The quantization noise will be shaped by this high pass filter, suppressing the noise within the signal's bandwidth.

3.3 Bandgap

The ADCs need a high performance voltage reference, with a temperature range of −40 to 125 °C. So the ultra low power reference structure introduced in the LDO module cannot meet the requirement. A traditional bandgap structure is used for this reference, with a typical performance of 10 ppm/°C.

The block diagram of this bandgap is shown in Fig. 7.

Fig. 7 Bandgap reference

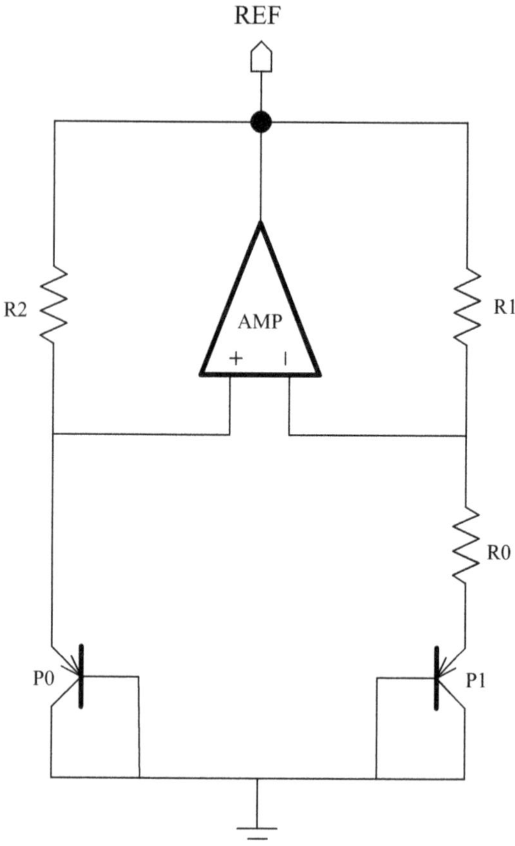

It's composed of two bipolar transistors, three resistors and one operational amplifier. The amplifier forces the voltages at the emitter of P0 and upper node of R0 to be the same. Suppose the emitter area of P1 is N times of P0, the resistor values of R1 and R2 are equal, then the voltage between R0 could be expressed as

$$\nabla V_{BE} = V_{BE,\,P0} - V_{BE,\,P1} = V_{T} \ln N \tag{2}$$

The output of this bandgap reference is

$$V_{REF} = \frac{R_1}{R_0} V_{T} \ln N + V_{BE,\,P0} \tag{3}$$

The first term of this expression is of positive temperature coefficient and the second term with negative temperature coefficient. By trimming the resistor value of $R1$, a reference with low temperature drift is generated.

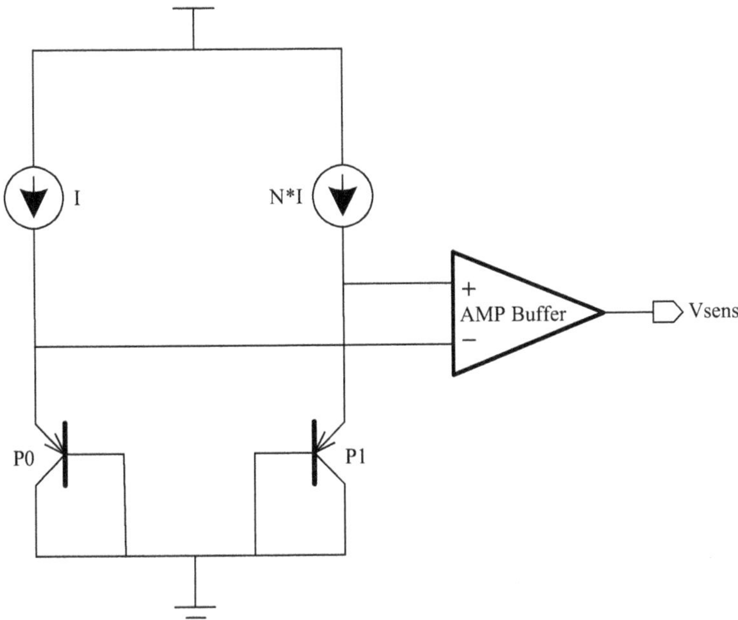

Fig. 8 Temperature sensor

3.4   Temperature Sensor

The temperature sensor enables the SoC to calibrate its RTC to compensate the frequency drift of the external crystal oscillator. The temperature sensor's principle is quite similar to the bandgap reference. It is actually a part of the bandgap. In the expression of bandgap output, the part $V_\mathrm{T}\ln N$ is a pretty linear curve versus temperature. The block diagram of the temperature sensor is shown in Fig. 8.

Unlike the bandgap reference, the emitter areas of these two bipolar transistors are the same, but the emitter current of P1 is N times of P0, so the expression is the same as the one in bandgap.

3.5   Ultra Low Power Crystal Oscillator

A 32,768 Hz crystal is needed to provide a stable clock for the SoC. The RTC module needs to be running even when the chip is in the sleep mode. The crystal oscillator needs to work in all situations. Therefore, the power consumption is key for this oscillator.

The block diagram of this crystal oscillator is shown in Fig. 9.

It's composed of one amplifier, two internal capacitors and three resistors. All the external components are integrated in this oscillator except the 32,768 Hz crys-

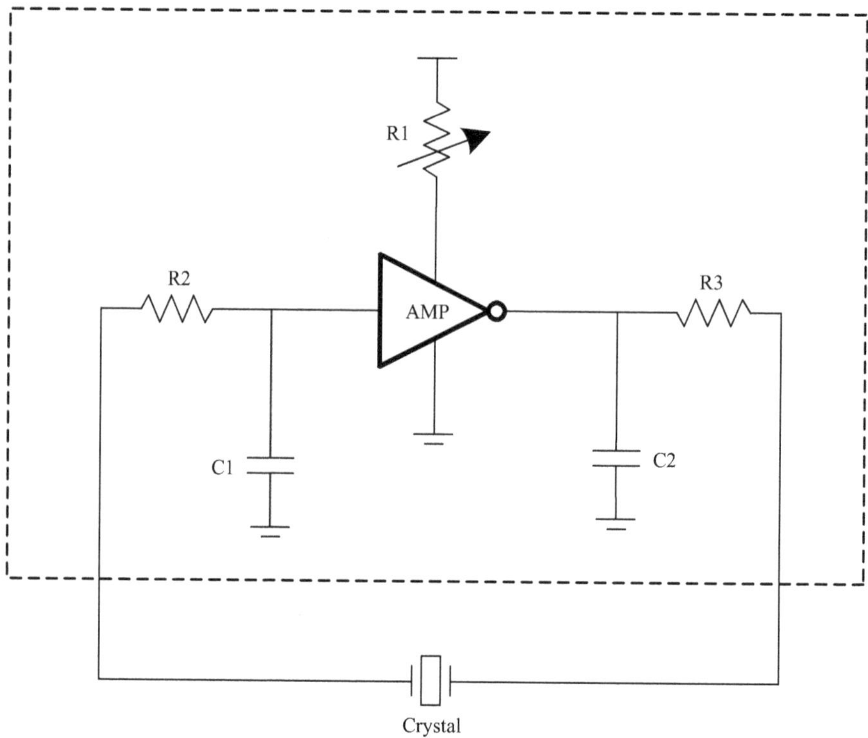

Fig. 9 Ultra Low power oscillator

tal, so there is no need to place two capacitors outside the chip. The resistance value
of resistor R1 can be tuned by the CPU to lower the power consumption of this
oscillator. The current consumption is less than 1 μA at normal status, and could be
decreased to 0.5 μA in deep sleep mode.

4 Instruction and Architecture Specific DSP

4.1 Computational Tasks

Figure 10 shows the signal flow diagram of the single phase energy metering algorithm. Each input channel has its own phase calibration, decimation, high-pass/
low-pass/band-pass, and averaging processing steps. The algorithm consists of filtering operations and a lot of multiplication, multiplication-addition, and square-
root operations.

In Fig. 10, PGA means Programmable Gain Amplifier, BPF means Band Pass
Filter, ITG stands for integral phase filter, AVG stands for averaging operations,
CNT means Counter, Energy Acc means Energy Accumulator.

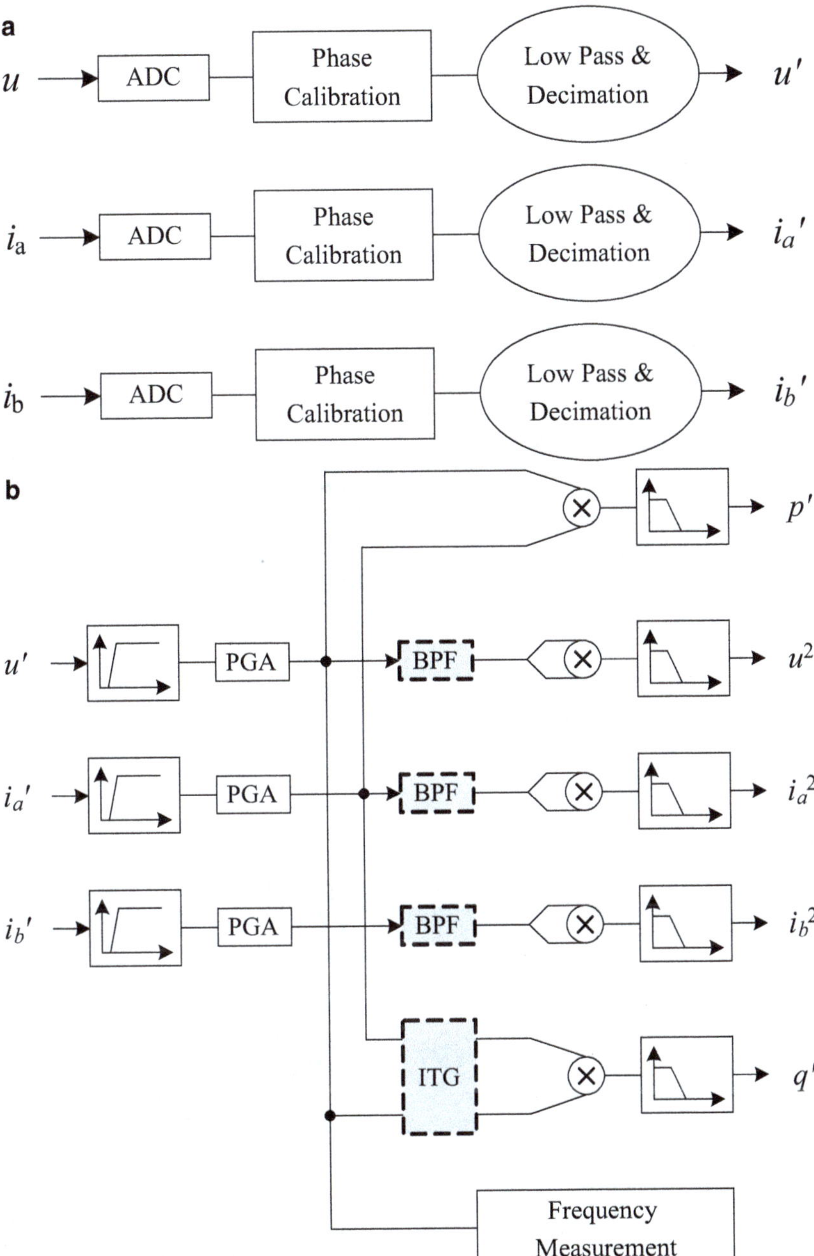

Fig. 10 Signal flow of single phase energy metering algorithm

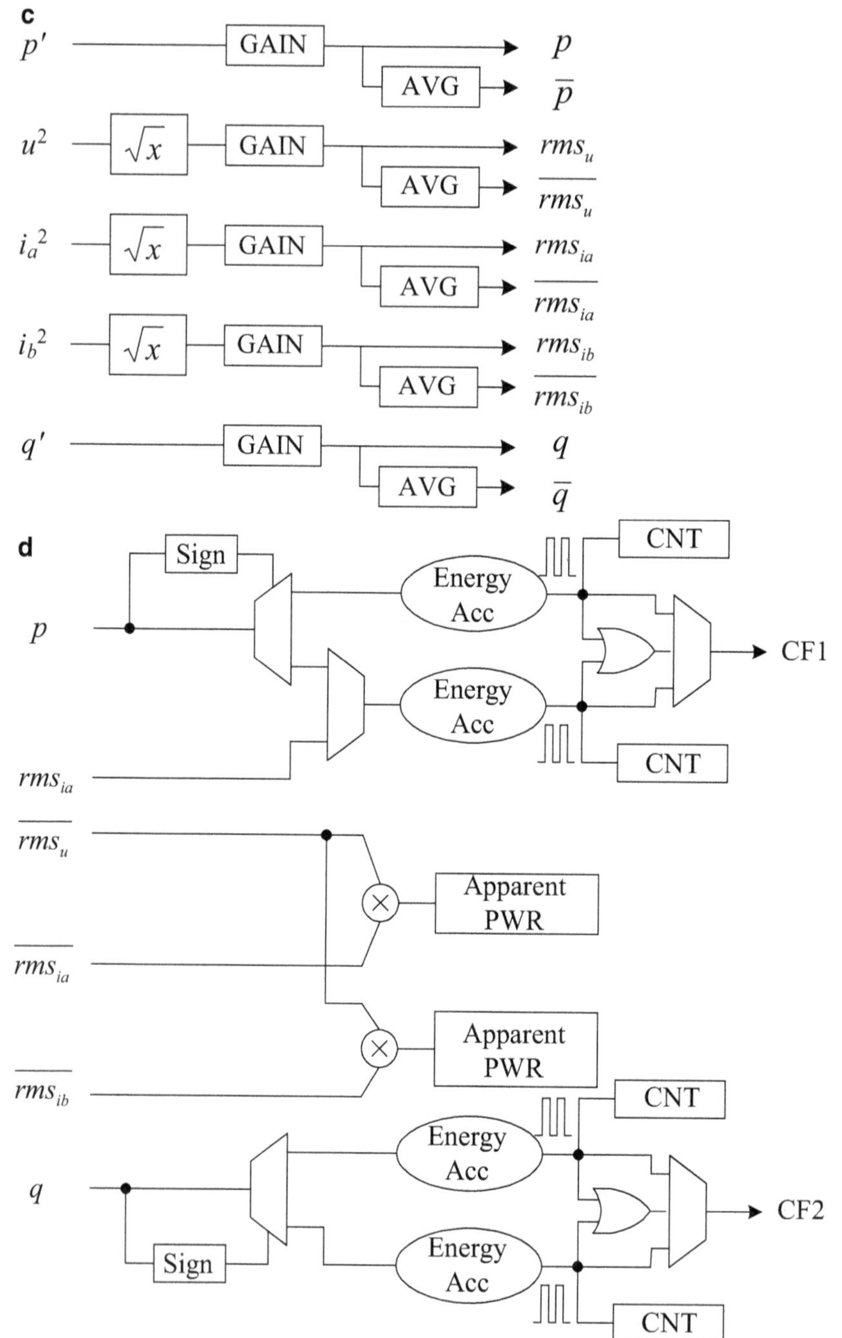

Fig. 10 (continued)

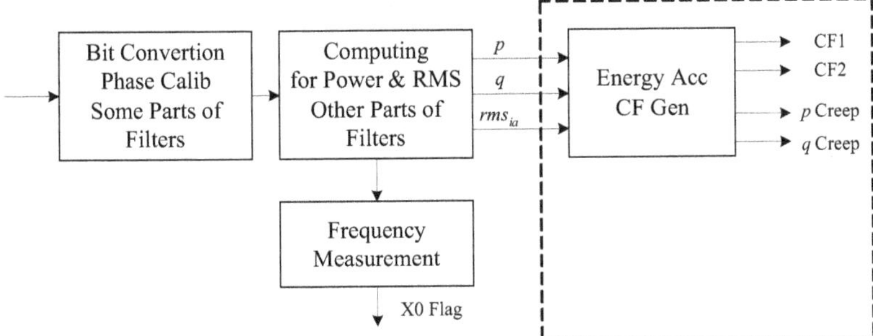

Fig. 11 Architecture diagram of DSP

Table 1 Instruction segments

Seg.	Bit20–Bit17	Bit16–Bit13	Bit12–Bit5	Bit4–Bit0
Func.	Instruction type	Operand register	Data address	Shifting flag and shifting bits

4.2 Instruction Set and Overall Architeture

According to the sampling rate of ADC and the final energy metering error requirement, most of the filtering algorithms can be implemented into multi-rate filters. Multi-rate filters mean that the first half and the second half of the filter run at different clock rates. All the filtering and computing operations can be classified into four parts according to their bit rate. Each part has not the same but similar or relative bit rates. In universal DSP solutions, all parts run at the same clock rate, and the 'multi-rate' is achieved by software conditional branching executions. Figure 11 shows the architecture diagram of DSP inside the SoC. Each part can run at different clock rate to save power.

The instruction set of the DSP is in RISC style. All instructions have the same length and one instruction is executed in one clock cycle. Each instruction has 21 bits and can be organized into four segments, shown in Table 1. By careful design, some coefficients of filters can be realized as a power of two, in other words, the co-efficients can be simply implemented by shifting operations. Some small truncating operations can also be realized by shifting. All instructions have a shifting segment to conduct the multiplication by two's integral power or small truncating operations.

Bit 20 ~ bit 17 is the instruction type segment, shown in Table 2. All instruction types are carefully designed to just fit the algorithm's computing demands. There are no branching instructions to simplify the instruction execution unit and instruction pipeline design. All branching actions are implemented by the PC generation unit according to a prearranged pattern.

A typical instruction segment can be described as follows, illustrated in Fig. 12.

- Load data from memory into register A and shift at the same time,
- Load data from memory into register B and shift at the same time,

Table 2 Instruction type

Instruction type	Functional description
NOP	No operation
RDM	Read out data memory
WRM	Store to data memory
ADD	Addition
SUB	Subtract
MUL	Multiply
MAC	Multiply and then addition
SQT	Square root
DIV	Divide
ABS	Get absolute value
SHF	U/Ia/Ib channel programmable gain
GTW	Truncation

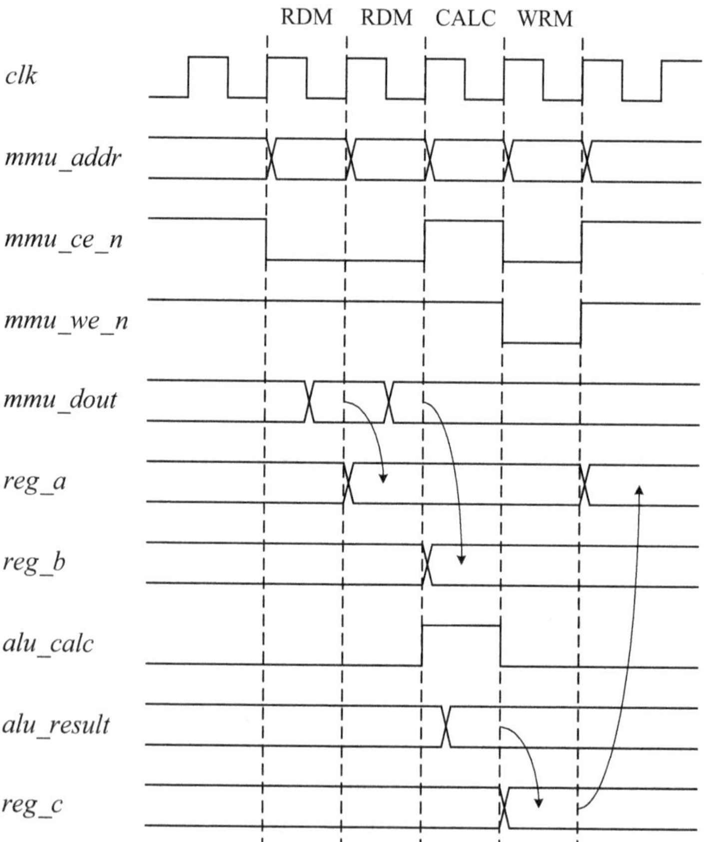

Fig. 12 Waveform of basic program segment

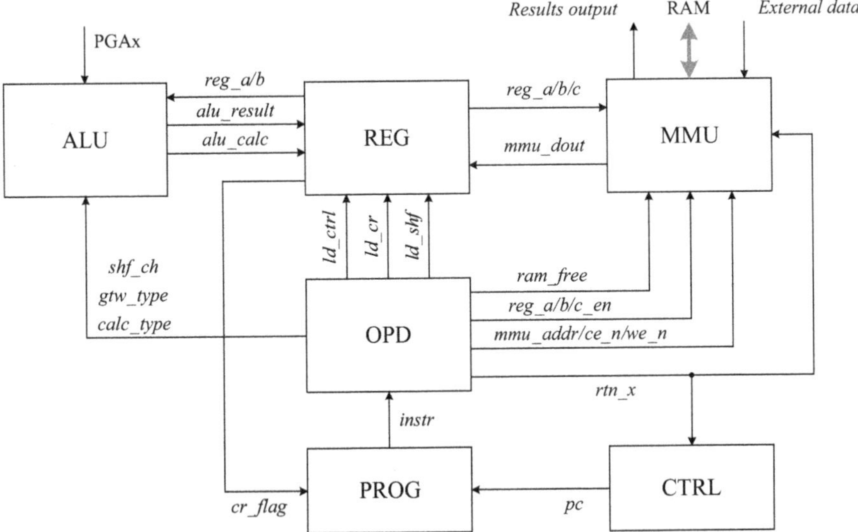

Fig. 13 Block diagram of the DSP

- Calculate A and B operands, shift the result at the same time, and save the shifted result into register C,
- Store register C into memory, update register A at the same time.

mmu_addr, *mmu_ce_n*, *mmu_we_n*, and *mmu_dout* are address input, chip enable, write enable, and output of the data memory respectively. The postfix *_n* means that the signal is electrically low level active. The data memory is active for reading or writing at each falling edge of clock signal to ensure setup and hold time requirements. *reg_a*, *reg_b*, and *reg_c* represent the values of register A, register B, and register C respectively. *alu_calc* is a signal indicating that the instruction currently decoded is one of computing type. *alu_result* is the calculation result. CALC represents any instruction of computing type.

The DSP is composed of opcode decoding unit OPD, register file unit REG, data memory management unit MMU, computing unit ALU, instruction fetching unit PROG, and run-time control unit CTRL. Figure 13 shows the block diagram.

The DSP is of Harvard architecture with independent instruction and data access channels. Controlled by computing type signals decoded by OPD, the ALU calculates with two operands output from REG. The computing result is loaded into REG. The REG loads data from MMU/ALU into register files and writes the contents of register files back into MMU. The MMU reads and writes data memory and external data, mapping them into different spaces of address. The CTRL generates program pointer *pc* according to program executing flag and the FSM (Finite State Machine) of multi-rate control. The PROG fetches instruction according to *pc* and sends *instr* to the OPD to be decoded. The hardware program branching is also realized in the PROG controlled by flag signals output from REG.

All the data memories are 32-bit wide. All registers and calculating resources in ALU are 64 bits to reduce truncation and round-off noise.

4.3 OPD

OPD stands for opcode decoder. The OPD decodes *instr* to get control or flag signals. The output of OPD can be classified into seven types.

Type 1; memory accessing control signals to MMU, including address signal *mmu_addr*, chip enable signal *mmu_ce_n*, write enable signal *mmu_we_n*, RAM accessing idle flag *ram_free. ram_free* is valid when the current instruction is not RDM or WRM.

Type 2; register selecting signals to MMU, including *reg_al_en, reg_ah_en, reg_bl_en, reg_bh_en, reg_cl_en, reg_ch_en*. These signals are used to select lower 32 bits of Register A, higher 32 bits of Register A, lower 32 bits of Register B, higher 32 bits of Register B, lower 32 bits of Register C, higher 32 bits of Register C to be loaded into MMU respectively.

Type 3; register loading control signals to REG, including *ld_al, ld_ah, ld_bl, ld_bh, ld_cl2al* and *ld_ch2ah*. These signals are used to control the process of loading MMU's output into lower half and higher half of Register A, loading MMU's output into lower half and higher half of Register B, loading MMU's output into lower half and higher half of Register C, and loading Register C's lower and higher half into lower and higher half of Register A respectively.

Type 4; shifting control signals to REG, including *ld_shf_op* and *ld_shf_bits*, to control the shifting procedure in data movement.

Type 5; sign bit latching control signal *ld_cr*. It is used to latch Register C's sign bit, ie. latching the sign bit of ALU's result.

Type 6; calculation control signals to ALU, including calculation type control signal *calc_add, calc_sub, calc_mul, calc_mac, calc_sqt, calc_div, calc_abs* and *calc_shf*; operand swapping control signal *sub_ba*; PGA control signal *shf_u, shf_ia* and *shf_ib*; data truncation control signals *gtw_x*, the postfix *_x* indicates different types of truncations in different filtering operations.

Type 7; program execution flags to CTRL and MMU, *rtn_x*, to control the generation of program pointer and lookup table pointer. They indicate that a program segment of some data rate has been executed. The control unit of program pointer should identify them and decide which address of the program code should be output next.

4.4 REG

REG stands for register files. There are two 64-bit operand registers, Register A and Register B, and one 64-bit computing result register, Register C, in the REG

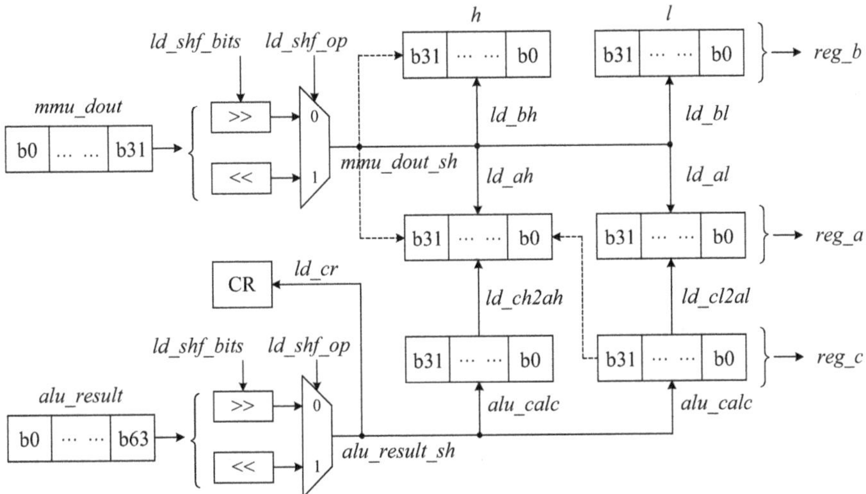

Fig. 14 Data flow and structure of REG

unit. There is also a sign bit register CR to save sign of ALU computing results. Figure 14 illustrates the data flow and structural blocks in REG.

The output of MMU, *mmu_dout*, has a bit width of 32. The output of ALU, *alu_result*, has a bit width of 64. Before being loaded into Register A/B/C, *mmu_dout* and *alu_result* are shifted according to *ld_shf_op* and *ld_shf_bits*.

Loading Register A and Register B can be divided into four operations, loading *reg_al* (lower half of Register A), loading *reg_ah* (higher half of Register A), loading *reg_bl* (lower half of Register B), loading *reg_bh* (higher half of Register B), activated when *ld_al*, *ld_ah*, *ld_bl* or *ld_bh* is in high level state. When loading *reg_al* and *reg_bl*, *reg_ah* and *reg_bh* will be stuffed by *mmu_dout_sh*'s sign bit.

When the current instruction is not NOP or RDM or WRM, *alu_calc* is active and *alu_result_sh* is loaded into Register C. The writeback operation of Register C to MMU can be divided into two parts, *reg_cl* (lower half of Register C) writeback and *reg_ch* (higher half of Register C) writeback. When *reg_cl* writeback is active, the *ld_cl2al* signal is valid and *reg_cl* is loaded into *reg_al* simultaneously, sign bit extension also valid in *reg_ah* according to *reg_cl*'s MSB. When *reg_ch* writeback is active, the *ld_ch2ah* signal is valid and *reg_ch* is loaded into *reg_ah* simultaneously.

alu_result_sh[63] is loaded into Register CR when *ld_cr* is active.

4.5 ALU

ALU stands for (algebraic) computation unit. The ALU performs calculations needed in filtering operations, such as addition, subtraction, multiplication, division,

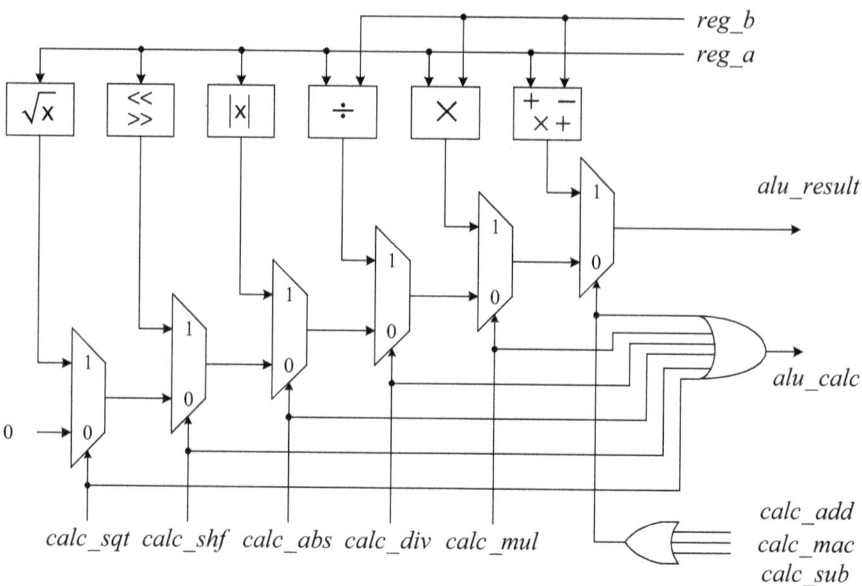

Fig. 15 Structure of ALU

square root, ABS, and PGA. The operand of square root, ABS and PGA is Register A. Other calculations have two operands, Register A and Register B. All calculation is conducted by independent hardware circuits in signed two's complement form. The paralleled results are selected by calculation type signals to be output as *alu_result*, illustrated in Fig. 15.

4.6 MMU

MMU stands for data memory management. The data address segment of the DSP's instruction set has a bit width of 8. Thus the maximum addressing ability is 256. Unified addressing mechanism is applied to MMU, *i. e.* all the data memory, constant filer coefficients, external data, and final results of the DSP are put into this 256 addressing space. The allocation of address is illustrated in Fig. 16.

4.7 PROG and CTRL

PROG and CTRL stand for instruction fetching unit and run-time control unit respectively. The CTRL generates *pc* according to *rtn_x* and a pre-set multi-rate calculation FSM. The smallest down-sampling cycle containing 1,024 clock cycles is defined as basic computing cycle. The computing tasks needing to execute every basic computing cycle are defined as fast computing. Other computing tasks

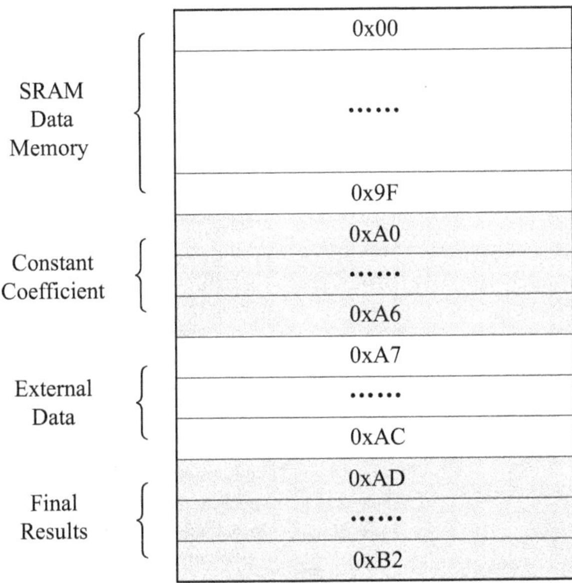

Fig. 16 Address mapping of MMU

needs executing every $1,024 \times 64$ clock cycles and every $1,024 \times 256$ clock cycles are defined as slow computing I and slow computing II respectively. If the slow computing is heavy, it can be partitioned into several basic computing cycles and combined with fast computing as a function. If not, the slow computing can be arranged directly in basic computing cycle. Since all computing tasks are carried out by the IAS-DSP in time division multiplexing, clock cycles that each function costs must be less than that of basic computing cycle.

Clock cycles needed by the IAS-DSP to finish each computing task are listed in Table 3. There is no task of more clock cycles consuming, so there is no need to partition them into shorter sub-tasks.

Combine fast computing and slow computing according to Table 4. The combined computing is defined as functions. Since func_0 is executed every basic computing cycle, the other functions must be finished in less than 686 clock cycles.

The iteration of slow computing I contains 16 basic computing cycles. The iteration of slow computing II contains 256 basic computing cycles. Circulating with 256 basic computing cycles, the functions inside each basic cycle is combined as listed in Table 5. A *rtn_x* is inserted after each function.

The multi-rate computing FSM controls CTRL to generate *pc*. The FSM has three states: *st_cycle_0* ~ *st_cycle_2*, according to three combining methods in Table 5. They are (func_0, *rtn_0*), (func_0, *rtn_0*, func_1, *rtn_1*), and (func_0, *rtn_0*, func_1, *rtn_1*, func_2, *rtn_2*). When each state accomplishes, the next state is entered according to 1,024 division, $1,024 \times 64$ division, or $1,024 \times 256$ division of the clock signal, as illustrated in Fig. 17. All fast computing updates data after *st_cycle_0* state accomplishes. Slow computing I updates data after *st_cycle_1* state finishes. When *st_cycle_2* state accomplishes, slow computing II updates its data.

Table 3 Clock cycles of each task

Type	No. of tasks	Notes	Clock cycles
Fast computing	Task_0	Nop	1
	Task_1	Slow part of low-pass decimation filters	39
	Task_2	Fast part of high-pass filters	60
	Task_3	DC compensations	12
	Task_4	PGAs	9
	Task_5	Integral phase filters	10
	Task_6	Band-pass filters	96
	Task_7	Multiplication for power and RMS	55
	Task_8	Fast part of low-pass filter for power and RMS	55
	Task_9	Output the sign of voltage channel	1
Slow computing I	Task_10	Slow part of high-pass filters	66
Slow computing II	Task_11	Slow part of low-pass filters for power and RMS	75
	Task_12	Non-linearity compensation of power	8
	Task_13	Square root operation for RMS	51
	Task_14	Gain operation of power and RMS	50
	Task_15	Apparent power calculations	22
	Task_16	Output active power, reactive power, RMS of two current channels	4

Table 4 Tasks combined into functions

No. of functions	Combining method	Notes	Clock cycles
func_0	task_0 ~ task_9	Fast computing	338
func_1	task_10	Slow computing I	66
func_2	task_11 ~ task_16	Slow computing II	210

In Fig. 17, each state is divided into two sub-states. In *st_cycle_0_a*, *pc* increases by one from the starting address of func_0 & func_0 till *rtn_0* is valid. Then *st_cycle_0_b* is entered, *pc* is forced back to &func_0 and kept stable till the 1,024 division of clock signal *clk_1k* is valid or the $1{,}024 \times 64$ division of clock signal *clk_1kx64* is valid, or $1{,}024 \times 256$ division of clock signal *clk_1kx256* is valid. Then *st_cycle_0_a*, or *st_cycle_1_a*, or *st_cycle_2_a* is entered according to *clk_1k*, *clk_1kx64*, and *clk_1kx256* respectively. In *st_cycle_0*, *pc* is generated according to Fig. 18.

In *st_cycle_1_a*, *pc* increases by one from the starting address of func_0 & func_0 till *rtn_1* is valid. Then *st_cycle_1_b* is entered, *pc* is forced back to &func_0 and kept stable till *clk_1k* is valid. Then *st_cycle_0_a* is entered. In *st_cycle_1*, *pc* is generated according to Fig. 19.

In *st_cycle_2_a*, *pc* increases by one from the starting address of func_0 & func_0 till *rtn_2* is valid. Then *st_cycle_2_b* is entered, *pc* is forced back to &func_0 and kept stable till *clk_1k* is valid. Then *st_cycle_0_a* is entered. In *st_cycle_2*, *pc* is generated according to Fig. 20.

The first line of instruction in func_0 is NOP. Forcing *pc* back to the starting address of func_0 and keeping it to be stable till the next basic computing cycle decrease the flip rate of circuits, thus lower power consumption of the IAS-DSP. According to Table 5, there are totally 262,144 clock cycles in 256 basic computing cycles, 87,263 clock cycles are occupied by calculation, the NOP clock cycles are 174,881. Measured in time, the active ratio of the IAS-DSP is only 33.288%.

Table 5 Functions combined in each basic computing cycle

No. of basic cycle	Combining method	Clock cycles
Cycle_0 ~ cycle_62	func_0, *rtn*_1	339
Cycle_63	func_0, *rtn*_1, func_1, *rtn*_1	406
Cycle_64 ~ cycle_126	func_0, *rtn*_1	339
Cycle_127	func_0, *rtn*_1, func_1, *rtn*_1	406
Cycle_128 ~ cycle_190	func_0, *rtn*_1	339
Cycle_191	func_0, *rtn*_1, func_1, *rtn*_1	406
Cycle_192 ~ cycle_254	func_0, *rtn*_1	339
Cycle_255	func_0, *rtn*_1, func_1, *rtn*_1, func_2, *rtn*_1	617

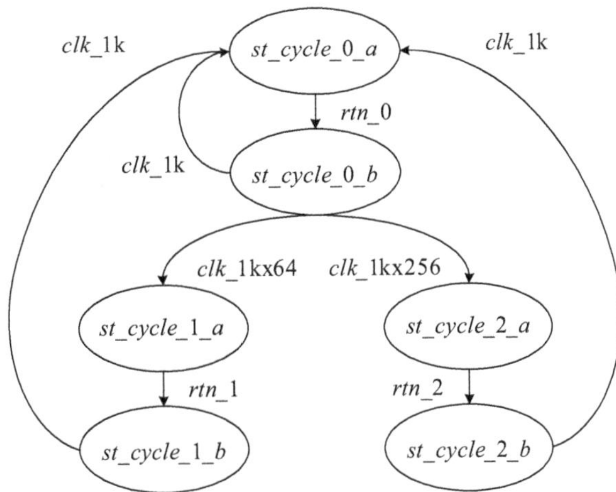

Fig. 17 The Multi-rate computing FSM

PROG has its instructions or programs arranged by func_0 ~ func_2, starting sequentially from address zero. PROG utilizes *pc* as its address to read a corresponding instruction out in hardware directly decoding way. The output is registered first, then sent to OPD. Thus a two-stage pipeline is realized. The IAS-DSP always does two operations simultaneously, fetching instruction and executing. Because of no branching instructions, there is no need to have abnormal-related mechanism in the pipeline, which simplifies hardware circuits.

5   Ultra Low Power Metering Modes

To satisfy critical power restrictive applications, several ultra low power metering modes have been designed. According to the power supply source, and the RMS values of voltage input and current input, the system control circuit gates some other circuits' clock or control the flip rate of D-flip-flops in that module to realize low power demands.

One of the SoC's input pins is used to detect the power supply conditions. The signal 'PWRUP' is defined to identify the present power supply source. If PWRUP is zero, the SoC is supplied by a battery. If PWRUP is one, the line supply is used.

Fig. 18 *pc* generation in *st_cycle_0*

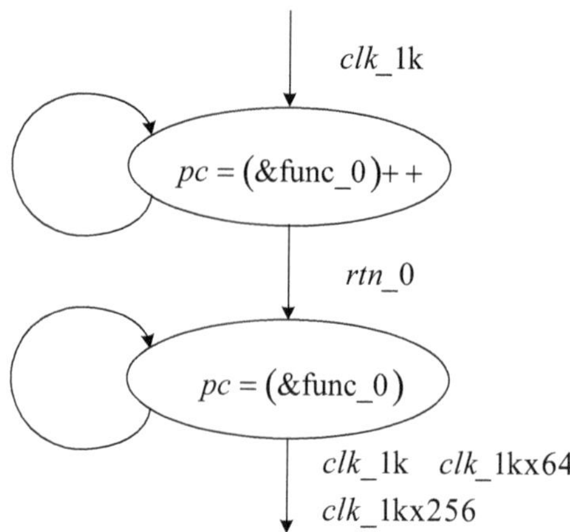

Fig. 19 *pc* generation in *st_cycle_1*

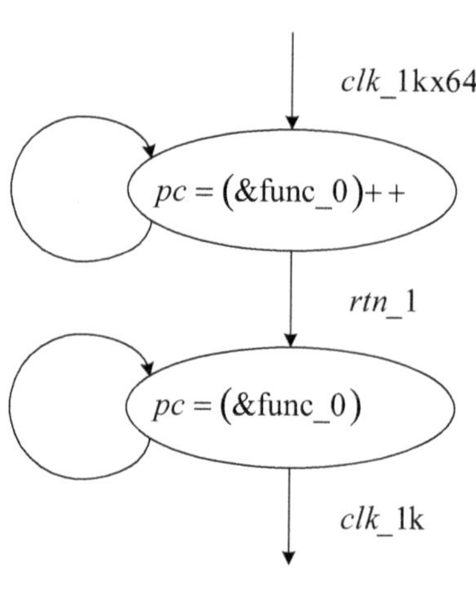

Fig. 20 *pc* generation in *st_cycle_2*

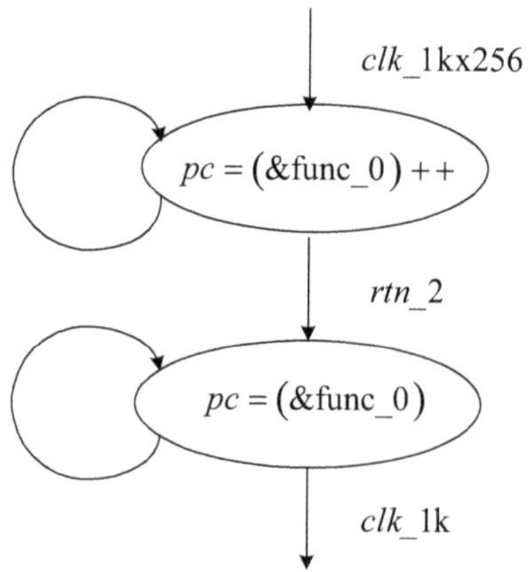

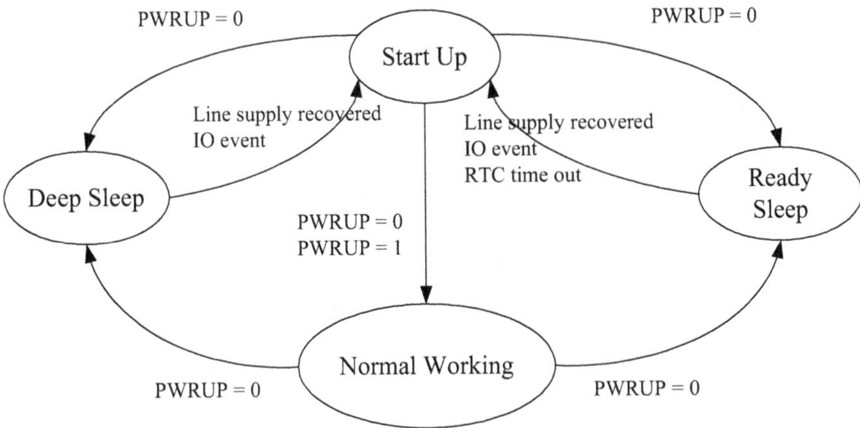

Fig. 21 Transitions of sleep states

5.1 Clock Domains and Their States

Clock Domains

There are two clock domains inside the SoC. The CPU core and its peripherals reside in domain 1 and the metering engine is in domain 2. Each of these two domains has its own independent clock source coming from the system control block which is controlled by software running on the CPU core. Clock domain 1 could use the 32,768 Hz clock (Low Frequency clock), or the $32,768 \times N$ (typical value of N is 100) Hz clock (High Frequency clock), or be clock gated. Clock domain 2 could use the 32,768 Hz clock (Low Frequency clock), or the 204,800 Hz clock (Reduced Frequency clock), or the 819,200 Hz clock (High Frequency clock), or be clock gated. After power up reset, both clock domain 1 and clock domain 2 use the 32,768 Hz low frequency clock.

When PWRUP signal is one, neither domain 1 nor domain 2 can be clock gated.

Ready Sleep and Deep Sleep State

If clock domain 1 is clock gated, the SoC is defined to be in Ready Sleep state or in Deep Sleep state. In Ready Sleep state, a transition from zero to one of PWRUP signal (from battery supply to line supply), an IO event (some input pin of the SoC has changed), or having been in Ready Sleep state for a certain time (RTC timing to some limit) could trigger a wake up reset to turn the SoC to the initial state. In Deep Sleep state, only a transition from zero to one of PWRUP signal and an IO event can trigger a wake up reset. The RTC timing event is disabled. The sleep states' transitions are illustrated in Fig. 21.

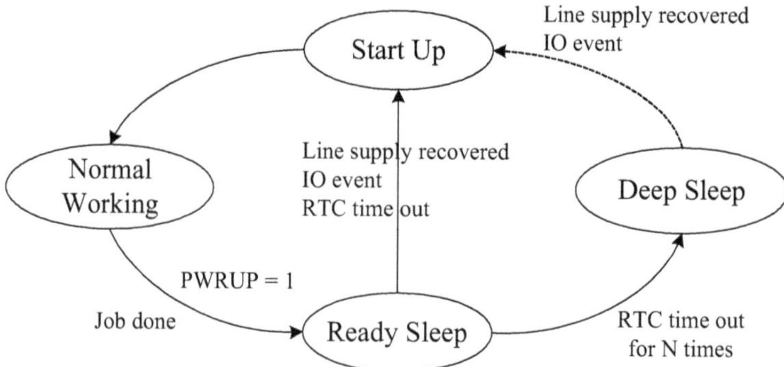

Fig. 22 State flow of keeping mode

Idle State

While the clock of domain 1 is active, all the inputs of the D flip-flops in domain 1 could be set to be stable or unchanged by some circuit techniques. This state of clock domain 1 is defined as Idle state.

5.2 Metering Modes

Keep Mode

Under the condition of PWRUP signal being zero, both domain 1 and domain 2 use the high frequency clock first. If the metering engine has both the RMS results of voltage input and current input lower than a preset threshold, the domain 2 is clock gated and the SoC goes into Ready Sleep state.

Wake up reset occurs after some time and the above procedure repeats. If the whole cycle counts to some limit, and the PWRUP signal never went to one nor the IO event occurred nor RMS values exceeded thresholds, the domain 2 is clock gated and the SoC goes into Deep Sleep state.

This is called the Keep Mode of the SoC, illustrated in Fig. 22. In this mode, only line supply recovery or an IO event can wake the SoC up. Otherwise the chip will be in Deep Sleep state consuming ultra low power supply. According to the fabrication technology, the current dissipation of digital circuits is typically hundreds of nano-amperes.

Reduced Frequency Mode

When some fault event happens and the voltage input is zero, the power line can not keep on supplying the SoC. The current transformer on board could generate induced voltage to supply it. But usually the current transformer has only a

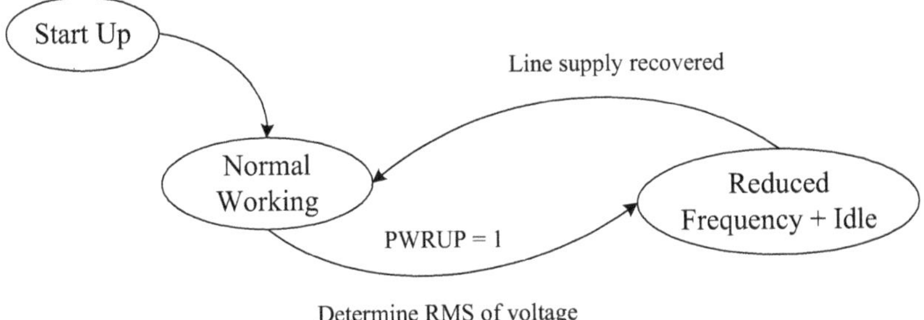

Fig. 23 State flow of reduced frequency mode

limited power capability which is smaller than the SoC's normal power consumption. The SoC must reduce its power consumption and maintain a certain metering accuracy.

If the metering engine detects that the RMS value of the voltage input is less than a certain threshold, then the SoC goes into Reduced Frequency Mode. In such conditions, the PWRUP signal is one, and neither domain 1 nor domain 2 can be clock gated. To reduce power consumption, domain 2 uses the reduced frequency clock, and domain 1 not only uses the $32,768 \times N$ (typical value of N is 25) Hz clock but also goes into Idle state when CPU is free of duty.

This is called the Reduced Frequency Mode of the SoC, illustrated in Fig. 23. In this mode, domain 2 keeps working but uses a slower clock. Circuits in domain 1 work intermittently, using a slower clock. The cost is less metering accuracy and less CPU horse power to handle tasks. But the current dissipation is reduced to typically less than 1 mA.

Ready Sleep plus Constant Metering Mode

Under some faulty conditions, the induced power by the current transformer is too small to supply the SoC chip, or there are no current transformers at all, or the fault condition is just a normal power failure, thus the SoC is totally supplied by battery. In that condition, the PWRUP signal is zero. Both domain 1 and domain 2 use high frequency clocks. If the metering engine detects that the RMS value of the voltage input is less than a certain threshold and the RMS value of the current input is greater than a certain threshold, the current RMS value is set to be the input of the energy accumulation circuit. Then clock domain 2 goes into Constant Metering state and the SoC goes into Ready Sleep state.

Wake up reset occurs after some time. The above procedure repeats. If the PWRUP signal never went to one nor the IO event occurred, the whole cycle will be repeated continuously. The state flow diagram is illustrated in Fig. 24. The current dissipation of SoC is typically about 50 μA.

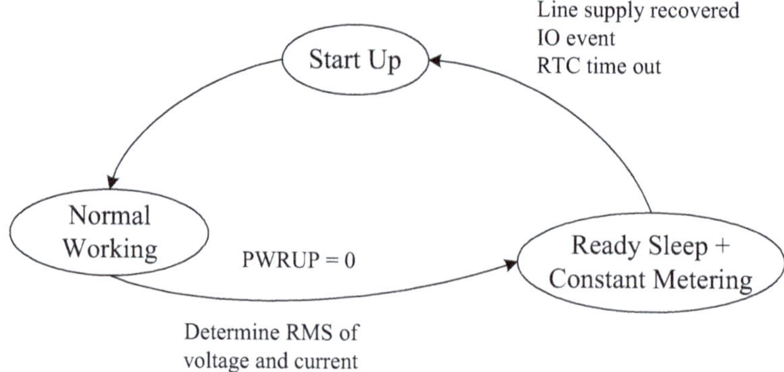

Fig. 24 State flow of ready sleep plus constant metering mode

Fig. 25 Photograph of the SoC

6   Experimental Results

The SoC chip was fabricated in TSMC 0.25 um mixed-mode embedded FLASH technology. The die micro photo is shown in Fig. 25.

The accuracy performance was measured by comparing the energy pulse output with the output of a standard energy meter sourced by the same standard power signal generator. In a dynamic input range of 1:5000, the SoC's metering error is less than 0.1%. This performance can be maintained from -40 to $85\,°C$ working conditions. EMI rejection features have been proved to have 15 KV contact and non-contact discharge tolerance.

Table 6 Current dissipation results

Item or Condition	Current dissipation
IAS-DSP in full speed	1.2 mA
SoC in normal mode	5 mA
SoC in keeping mode	5 μA
SoC in reduced frequency	0.9~1 mA
SoC in ready sleep plus constant metering	50 μA

Some current dissipation results were measured and listed in Table 6. The current dissipation of IAS-DSP is comparable with the mainstream energy metering front ends fabricated in 0.18 or 0.13 μm process. In Keep Mode, the RTC, LDO, and temperature error compensation circuit still work. The 50 μA current dissipation result of Ready Sleep plus Constant Metering can significantly improve the ability to keep metering in condition of zero line fault.

References

1. Dewen Wang, Yaqi Song, Yongli Zhu, "Information Platform of Smart Grid Based on Cloud Computing," Automation of Electric Power System, vol. 34, no. 22, pp. 7–12, Nov. 25, 2010.
2. Long Zhou, Fangyuan Xu, Yingnan Ma, "Impact of Smart Metering on Energy Efficiency," Proc. of the 9th International Conf. on Machine Learning and Cybernetics, vol. 6, pp. 3213–3218, July 2010.
3. Chih-hsien Kung, M.J. Devaney, "Multirate Digital Power Metering," Proc. of the Instrumentation and Measurement Technology Conference, 1995.
4. Yongjun Feng, Zhiyong Pu, Zisong Jiang, "Design of 1(100) A Single Phase High Accuracy Energy Meter Using 5,000:1 Dynamic Range Metering IC," Electrical Measurement & Instrumentation, vol. 48, no. 541, pp. 50–53, Jan. 2011.
5. Zhigang Hu, Kai Xu, Yongfeng Cui, "Application of Energy Measurement IC ADE7878 in Watt-hour Meter," Electrical Measurement & Instrumentation, vol. 47, no. 7A, pp. 128–131, July. 2010.
6. Yan Zhao, Nianxiong Tan, et. al, "A Single-Phase Energy Metering SoC with IAS-DSP and Ultra Low Power Metering Mode," Proceedings of IEEE International SoC Conference, pp. 354–358, September 2011.

SoC for Hearing Aids

Kun Yang, Liyuan Liu, Shupeng Zhong and Nianxiong Nick Tan

1 Introduction to Hearing Aids ICs

Hearing aids design has changed dramatically since the end of the 1980s, before which only analog hearing instruments that can only amplify sound were used.

Figure 1 shows a typical analog hearing aids device. It consists of a microphone, a pre-amplifier, a tone control, an amplifier, and an earphone. An acoustic input signal is converted to an electronic input signal by the microphone. The preamplifier amplifies this electronic input signal. The frequency response of this electronic input signal is shaped by the tone control. After this shaping, the amplifier again amplifies the signal. Finally the signal is converted back to an acoustic output signal by the earphone. Both the acoustic and electronic signals are processed in analog world.

Later came some analog hearing aids devices that are digitally programmable. Some parameters of the analog components, such as tone control and amplifier, can be stored in a digital storage unit (any kind of memory) and can be adjusted for different users.

In contrast to an analog hearing aids device, a digital device has an analog-to-digital converter (ADC), a digital signal processor (DSP), and a digital-to-analog converter (DAC). The amplified analog electronic signals are converted into digital signals first, and then are processed by the DSP and finally converted back to

K. Yang (✉) · S. Zhong
Vango Technologies, Inc., Hangzhou, China
e-mail: yangk@vangotech.com

S. Zhong
e-mail: zhongsp@vangotech.com

L. Liu
Institute of Semiconductors, Chinese Academy of Sciences, Beijing, China

N. N. Tan
Institute of VLSI Design, Zhejiang University, Hangzhou, China

N. N. Tan et al. (eds.), *Ultra-Low Power Integrated Circuit Design*,
Analog Circuits and Signal Processing 85, DOI 10.1007/978-1-4419-9973-3_9,
© Springer Science+Business Media New York 2014

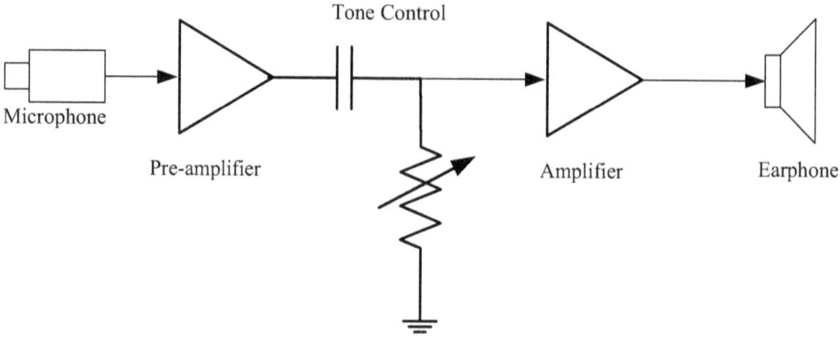

Fig. 1 Typical analog hearing aids device

analog electronic signals again. The introduction of the DSP realizes technology that makes digital hearing aids device very successful: multiband compression, directional microphones to avoid noise issues; and feedback cancellation that allows greater gain (Fig. 2).

Digital signal processing has many advantages over analog signal processing, including performance, miniaturization, reproducibility, stability, programmability, and signal-processing complexity. Hearing aids devices are available in different sizes and shapes. The first available portable hearing aids device is of pocket-size. Nowadays, most hearing aids device are behind-the-ear (BTE), or in-the-ear (ITE) devices. Hearing aids IC must be as small as possible to be integrated into these BTE/ITE devices. And the power consumption of hearing aids IC must be as low as possible because there is little space for battery in such devices.

2 Architecture of the Hearing Aids SoC

2.1 *Overall Architecture*

The block diagram of a digital hearing aids IC is shown in the following fig. 3. Analog components that are integrated into hearing aids include: ADC, Power Management, Clock Generator, Power On Reset, Reference, and Class-D amplifier as DAC. Digital part consists of COMB filter for ADC and interpolation filter for DAC, DSP core, system control, I-SRAM (static random access memory) for instruction and D-SRAM for data, I2C, JTAG (Joint Test Action Group) and general purpose input output (GPIOs), etc.

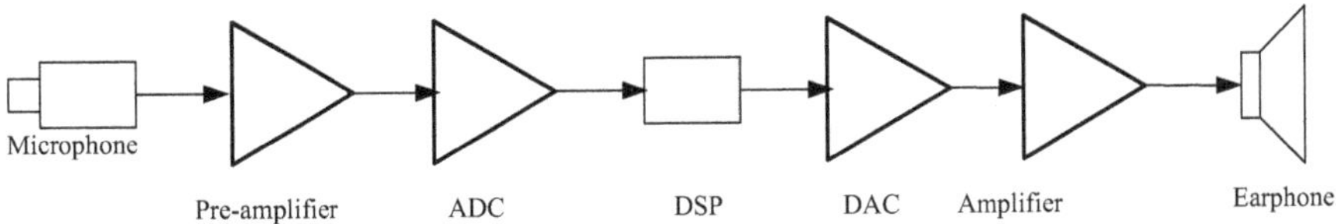

Fig. 2 Typical digital hearing aids device

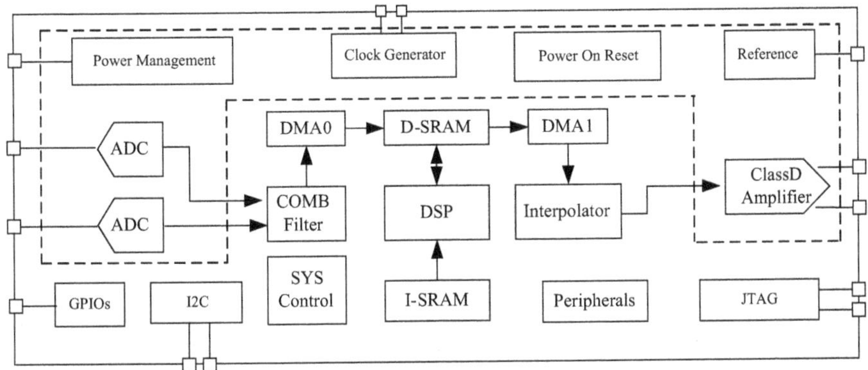

Fig. 3 Architecture diagram of hearing aids SoC

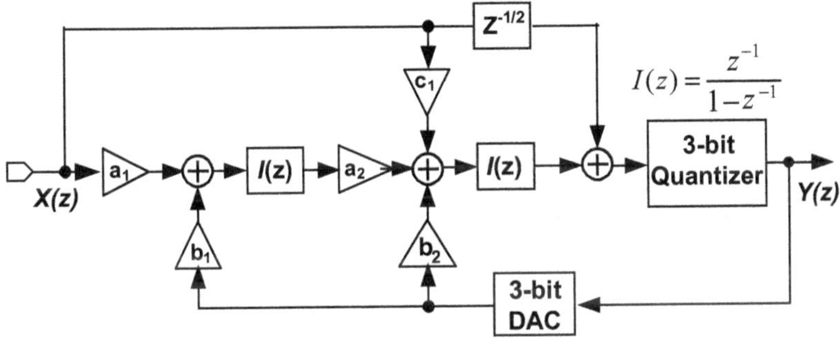

Fig. 4 Delta-sigma modulator

2.2 ADC

Two channels of high precision delta-sigma ADC are integrated into hearing aids SoC
to support directional microphones. Delta-sigma ADC consists of three units, an anti-
aliasing filter (a low-pass filter) is a just simple RC filter, a delta-sigma modulator, and
a decimator filter. Figure 4 shows the modulator architecture. Second order modulator
is adopted. The oversampling ratio (OSR) is chosen as 128 for sufficient low quantization error. The coefficients a1, a2, b1, b2, c1 are chosen as 7/11, 6/8, 7/11, 7/8, 7/8
respectively. The signal transfer function has all pass characteristic. The feed-forward

path helps remove the input signal related component from the output swing of the integrators and only quantization error signal exists. By using a 3-bit quantizer, the output swings of the 1st stage and the 2nd stage integrators can be further reduced. Simulation shows that swings are within ± 0.18 V and ± 0.25 V respectively. In traditional design, the quantizer samples the signal during $\Phi 1$ and makes decision during $\Phi 2$. The dynamic element matching (DEM) and integration phase also take place in $\Phi 2$, which requires high speed comparator and DEM. An improved timing is proposed in this design. That is the quantizer samples at integration phase $\Phi 2$. So quantization and DEM can occupy the whole of $\Phi 1$. This slightly change of timing relaxes the speed requirement of comparator and DEM. Although this modification will add an additional half period delay in the feed forward path, the transfer function will not be changed significantly because of oversampling.

Figure 5 shows the switched-capacitor circuits of the modulator. Two stages share the capacitor network that reduces circuits' complexity. For the 16 kHz signal bandwidth, the sampling rate is 4.096 MHz. The amplifier in the first integrator is chopped to remove flicker noise from low frequency range. To let the chopping happens in the middle of $\Phi 1$, a faster clock is introduced. The chop frequency is chosen as 128 kHz in order that the flicker noise left in the signal band is small enough. The sampling capacitance is chosen to be 2pF for low thermal noise. The output of the 2nd integrator is sampled and quantized into 3-bit digital codes by a flash quantizer. The references are generated using a resistor ladder. The digital codes are firstly processed by DEM logic and then feedback to the modulator loop through capacitance DAC. The using of 3-bit quantizer also enhances the modulator stability and guarantees that the modulator has an overload level of 0.9.

The block diagram of the decimation filter is shown in Fig. 6. The decimator is comprised by five stages comb filters cascaded by two stages of half-band filters. The multi-stage design relaxes every sub-filter design. The input signal is down sampled by a factor of 2. The comb filter has transfer function of $(1+z\text{-}1)4$. Comb filters and half-band filters adopt poly-phase topology. With sampling rate lowing down, filter in every stage needs to operate at a slower rate than the sampling rate fs. Especially the half band filters in the last two stages, they only need to work under fs/64 and fs/128. We choose fs as master clock for each stage and thereby computation such as accumulate or shift can be distributed in time by multiplexing the basic arithmetic unit.

2.3 DAC

The DAC in hearing aids SoC consists of three parts: an interpolator, a delta-sigma modulator, and a class D amplifier for high output power efficiency. It is shown in the following figure (Fig. 7).

The oversampling ratio (OSR) is set to 64 based on calculation and simulation, which will fulfill the performance requirement. To avoid big exponent number, we design the interpolation filter in the way of cascade of several stages, two stages of IIR filter and four stages of comb filter, rather than a single stage, shown in Fig. 3.

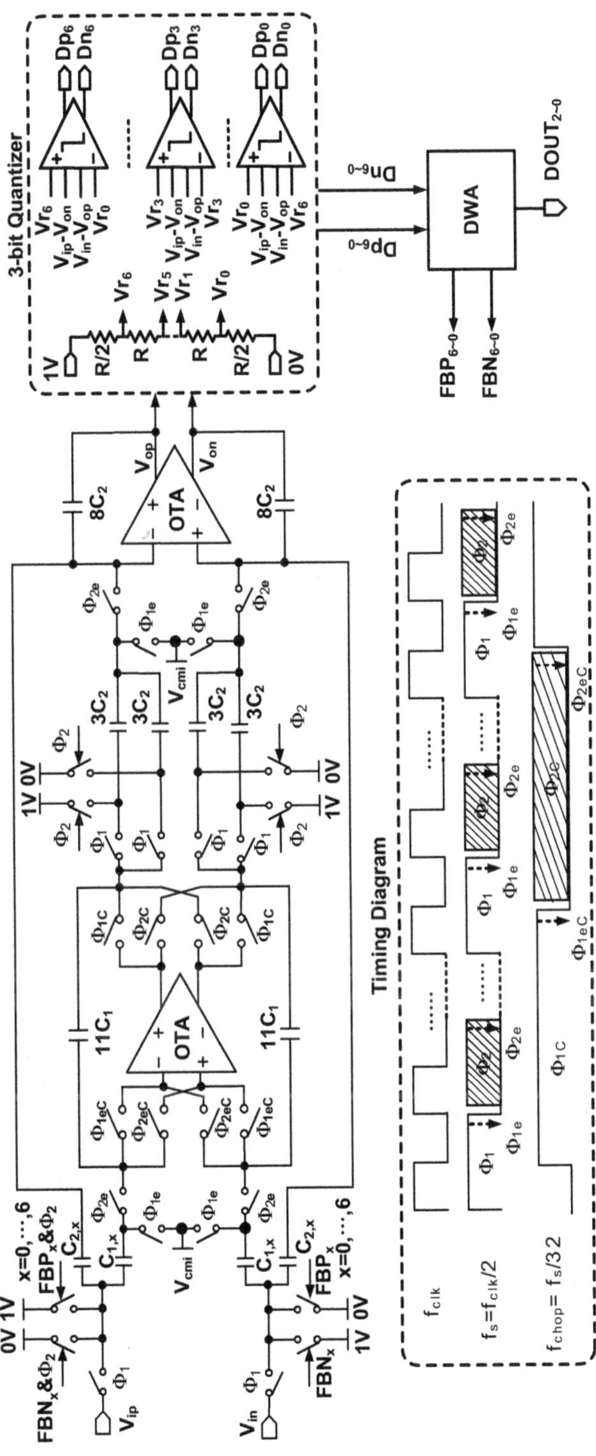

Fig. 5 SC realization of the delta-sigma modulator

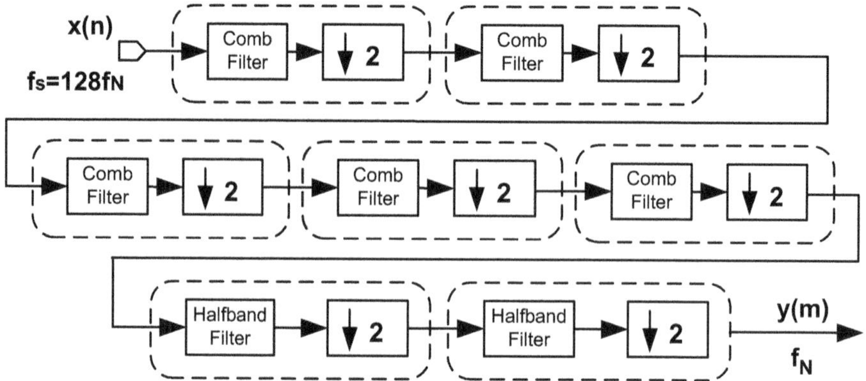

Fig. 6 Decimation filter for the delta-sigma modulator

Fig. 7 DAC for the hearing aids

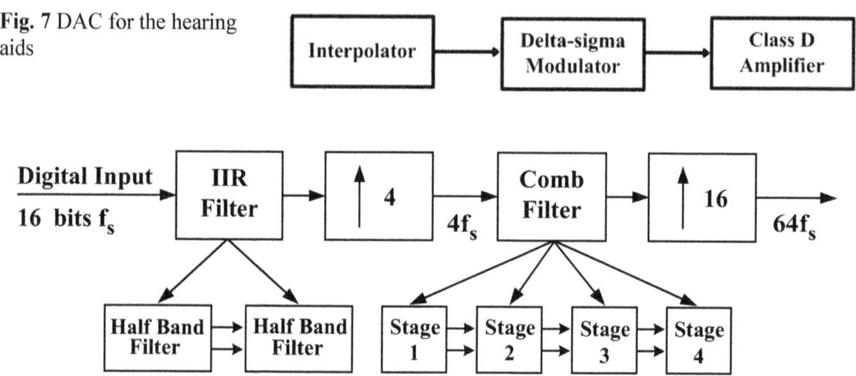

Fig. 8 Interpolator

Every stage realizes two times of oversampling. Thus the total interpolator realizes an oversampling ratio (OSR) of 64 as shown in the following figure (Fig. 8).

The interpolation filter consists of a comb filter followed by 2 half-band filters similar to the decimation filter. The delta-sigma demodulator converts the digital word into one bit stream before driving the class amplifier. In order to reduce the noise, 1.5 bits instead of 1 bit truncation is used. Most design implementations choose FIR filter rather than an IIR filter, because of the strict requirements on the linear phase characteristics in audio applications. Our design seeks a tradeoff between the circuit complexity and the linear-phase, so an IIR filter with an approximate linear phase which can use a small exponent number is used in this design. Half-band filtering techniques are efficient to reduce the circuit complexity in design of IIR filters, because half of the coefficient numbers are zero-value. The system clock is $64/2^N$ times of the Nth stage working frequency because of oversampling, so there are $64/2^N$ time-slots can be used for the Nth stage. Time division multiplexing technology is used in the design of interpolation filter. We use a hardware-efficient method of complex sequential control logic (BMCU) to realize the IIR filter in half band

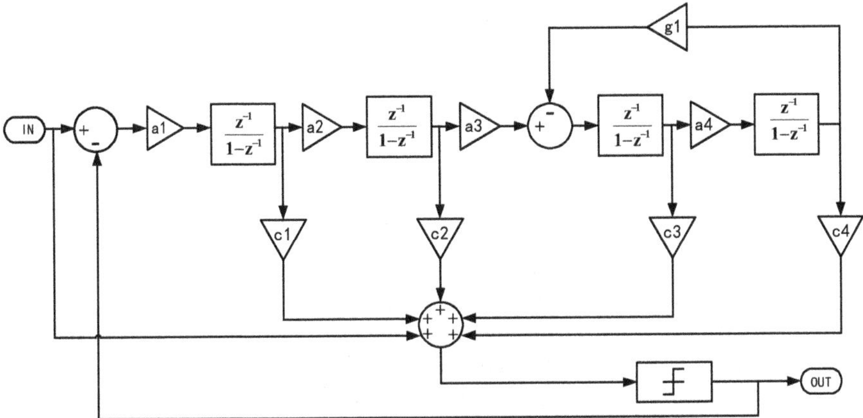

Fig. 9 Delta-sigma demodulator

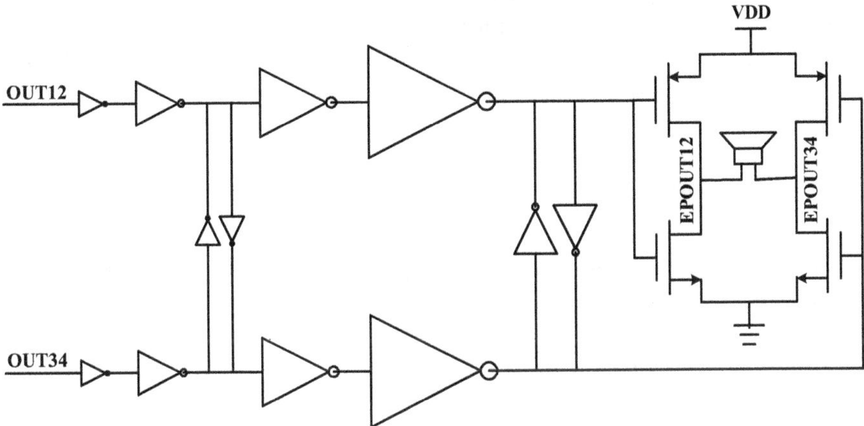

Fig. 10 Proposed class D output stage

structure. A multiplexing cell which can be reused several times is shown in the following figure. The comb filter in half band structure is realized according to the same principle.

CIFF structure with local resonator feedback around pairs of integrators in the loop filter is used to implement the delta-sigma modulator. A notch in the spectrum caused by the resonator is shown in Fig. 9 (considering a modulated 20 bits sine wave whose frequency is 3,750 Hz, amp is 0.8 V), which will realize a better result of noise shaping and reach a higher SNDR. A dither is added to the resonator before the quantizer of the $\Delta\Sigma$ modulator to suppress the idle tones. It is generated using a linear feedback shift register (LFSR) in Galois configuration. The generator polynomial of the LFSR is $x^{19} + x^{18} + x^{17} + x^{14} + 1$.

Typically, the 1-bit output of the $\Delta\Sigma$ modulator will control the switching activity of the class D output stage directory. A DLPF encoder is placed after the $\Delta\Sigma$ modulator, whose transfer function is $H(z) = \frac{1}{2}(1 + z^{-1})$, shown in Fig. 10. It will result in a three state output $(-1, 0, 1)$. By using such an encoder, the output data stream is composed mostly by 0s. This can be explained as that H(z) servers as a low pass filter which remove high frequency quantization noise at fs/2 and results into lower transition rates in the output data streams.

In order to get a higher output, the full-bridge topology is adopted as the power output stage, which consists of two sets of switches. A full-bridge topology is suitable for open loop design, and the different output structure of the bridge topology inherently can cancel the even order of harmonic distortion components and DC offsets. Figure 10 shows the hardware configuration of the proposed class D output stage.

Dead-time control is very important for a class D amplifier, because once both high and low side MOSFETs are turned on simultaneously, a low resistance between power supply and ground will result in a large current, which can cause significant energy loss and a crossover-type distortion or even make the MOS be shoot though. In this design we should ensure that the signals in each branch of the differential path have equal rise and fall times, weak cross-coupled inverters are inserted between the two differential lines of the output stage. The feedforward provided by the cross-coupled inverters helps minimizing the signal skew.

2.4 Other Analog Units

There are some other analog units that supports ADC and digital circuit to work right.

BGP

The Band-Gap circuit supplies reference voltage and current to ADC/DAC. This Band-Gap circuit can output a reference voltage of about 1.12 V which is affected little over the temperature variation, with a typical temperature coefficient of 20 ppm/°C. Enable the Band-Gap circuit, and then, enable the other circuits. The temperature coefficient can be adjusted for adjustment for production variation.

Power Management

Power management circuit consists of 4 low-dropout regulators (LDO) for analog/digital/IO and class-D amplifier. The LDO has the following features:

- Ultra low power that the LDO itself only consumes 0.2 μA current;
- LDO Outputs a stable voltage when voltage of input power pin is 200 mV higher than the output voltage;

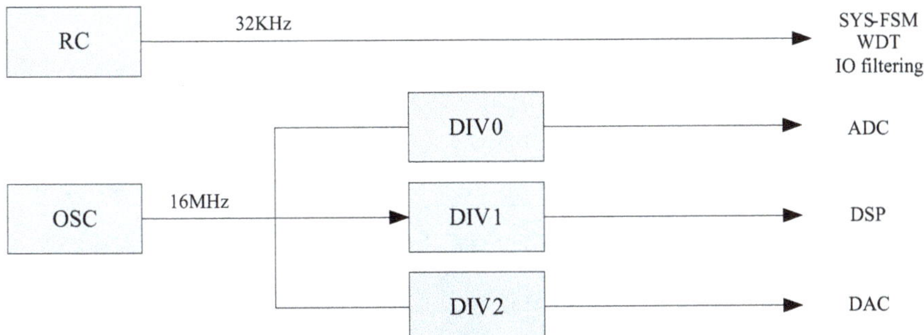

Fig. 11 Diagram of clock generator

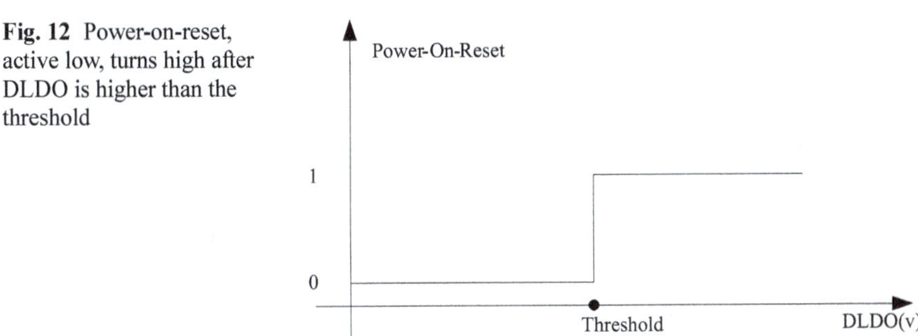

Fig. 12 Power-on-reset, active low, turns high after DLDO is higher than the threshold

- Decoupling capacitors are necessary to connect off-chip to the LDO, 4.7 µF and 0.1 µF capacitors in parallel are recommended;
- LDO output is programmable for better power consumption/performance trade-off.

Clock Generator

Clock generator includes 2 clock sources: resistance and capacitor (RC) clock, generated by the on-chip RC oscillator circuit; OSC clock, generated by the crystal oscillator circuit, composed of an external 16 MHz crystal and an on-chip oscillator circuit. As shown in Fig. 11, the RC clock is used as source for system finite state machine (FSM) management, watch dog timer (WDT) and input signal filtering; OSC clock is divided to 3 clock sources for ADC/DAC and DSP.

Reset

A reliable POR (power on reset) is generated to reset all units in SoC.

In hearing aids SoC, there are 2 reset sources ensuring the proper reset: the power-on reset, providing the power-on reset circuit, and the external reset pin input.

Figure 12 shows that the power-on reset circuit monitors the output voltage of DLDO (Digital-LDO) for digital part. When DLDO output voltage is lower than the

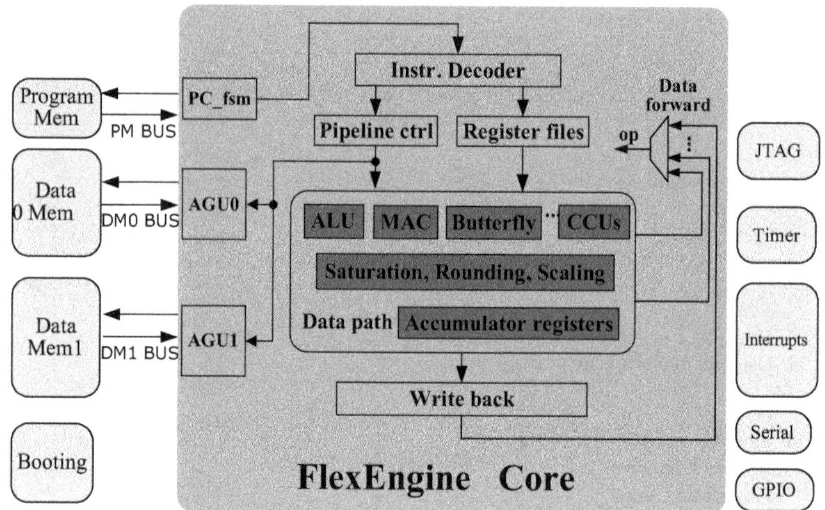

Fig. 13 FlexEngine DSP architecture

threshold, for example, 0.8 V, determined by the chip, the system stays in the reset state, and when it is higher than the threshold, the power-on reset signal is released, and the system exits the reset state.

The input signal on the external reset pin was filtered via the RC clock to avoid static disturbance. This signal must be driven low for at least 5 ms.

2.5 DSP Core

An ultra low power application-specific instruction-set processer (ASIP) called FlexEngine is integrated into hearing aids SoC for audio application, such as audio encoding and decoding, sound enhancement algorithms, noise reduction. The FlexEngine is a 24-bit low power DSP core, which has optimized instruction set and high-efficient microarchitecture. As FlexEngine DSP is a well balanced core, achieving low power, high efficiency, and low gate count, many audio domain implementations can benefit from it.

The microarchitecture of FlexEngine DSP is illustrated in Fig. 13. The processor core introduces distributed Harvard memory architecture, using one program memory and two separate data memories. The pipeline scheme partitions each instruction into 5 sub operations, achieving parallel execution for each cycle. The critical path for each pipeline stage is well balanced to gain a maximum clock frequency. Customized instructions are added to accelerate the audio processing. Meanwhile, the hardware of the accelerator is carefully designed with low power and low gate count. Zero overlapped hardware loops are developed to eliminate the subroutine jump overhead. The Address Generation Units (AGUs) offer extensive addressing mode for regular embedded signal process. The advanced data forward scheme im-

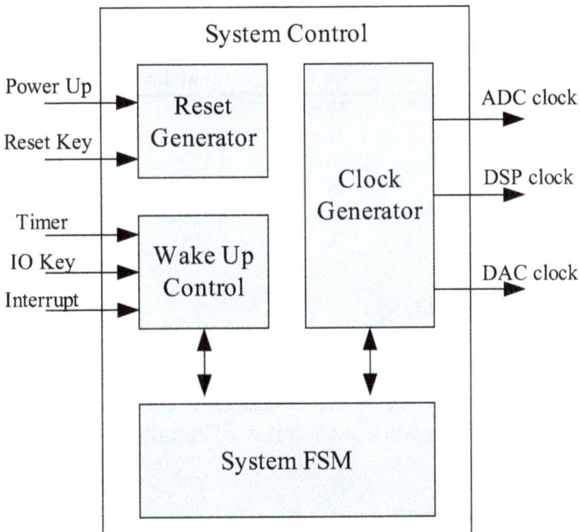

Fig. 14 System control diagram

proves the execution parallelization by hiding various data hazards. The data path consists of arithmetic-logic unit (ALU), Multiply-accumulate (MAC) and customized computation units.

Hardware features include:

- Distributed harvard memory architecture;
- Full 24-bit data paths, 64 K words address space for program and data memories;
- Reduced Instruction Set Computer (RISC) instruction set, suitable for both control and computation;
- Zero overhead loops;
- Various addressing modes with modulo protection, bit reversal;
- Data forward scheme;
- Four 56-bit accumulators;
- Supporting saturation, rounding and scaling operations;
- Customized computation units for audio acceleration;
- JTAG: breakpoint, single step, run/stop, memory read/write.

2.6 *System Control Circuit*

The system control block implements the SoC's power strategy, realizing boot-loader and debug function. The block diagram is shown in the following figure (Fig. 14).

Reset generator, clock generator, wake up control and a system FSM is integrated in the system control unit.

Table 1 Clock frequency strategy

Bandwidth (in K)	Data rate (in K)	ADC clock (in MHz)	DAC clock (in MHz)	DSP clock (in MHz)	System clock (in KHz)
16	32	8.192	2.048	16.384	32
12	24	6.144	1.536	12.288	32
10	20	5.120	1.280	10.240	32
8	16	4.096	1.024	8.192	32
6	12	3.072	0.768	6.144	32
4	8	2.048	0.512	4.096	32

Power Strategy

System control circuit implements SoC's power strategy majorly through adjusting working frequency of ADC and DSP core to achieve the best trade-off between performance/power consumption of certain situation. Performance includes the following items:

- ADC accuracy
- Signal band-width
- Complexity of DSP algorithm (Table 1)

System control circuit also introduces IDLE mode for conditions that no audio processing is necessary. Through clock gating major digital units and powering down most analog circuit, hearing aids SoC can achieve a power consumption of several µA in the IDEL mode. In IDLE mode, only system control keep working on an internal RC clock whose frequency is about 32 KHz.

Communication/RTC/wakeup pin and reset events can wake up SoC back to work mode.

The FSM of the SoC work mode is show in Fig. 15.

Boot-Loader

All firmware is stored in an off-chip memory. The boot-loader automatically loads all firmware from off-chip memory and stores to internal instruction SRAM after every reset of the chip. The block diagram is shown in the following figure (Fig. 16).

Boot-load includes an I2C master to access off-chip memory.

Firmware program and update are achieved through this I2C master.

After all firmware is moved to on-chip instruction memory, the off chip memory is shut down to save power.

On-Chip-Debugger (OCD)

A small OCD is implemented into hearing aids SoC for SRAM & peripheral access. It has the following functions:

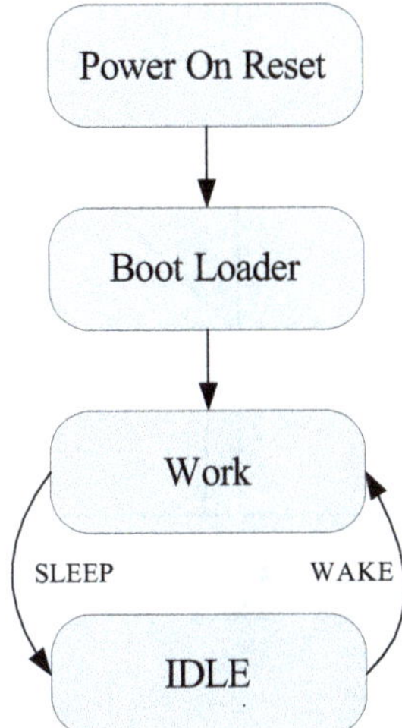

Fig. 15 System FSM

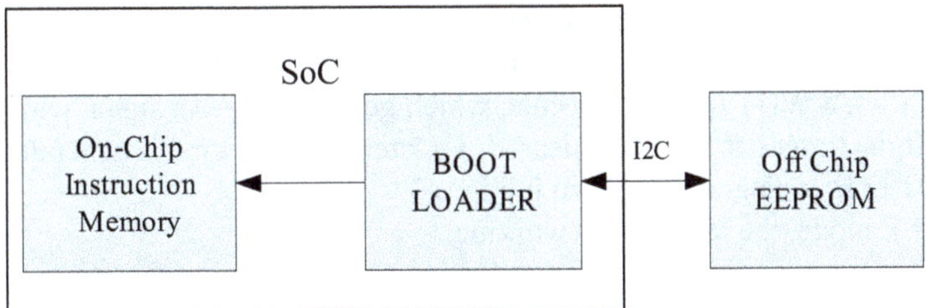

Fig. 16 Boot-loader of hearing aids SoC

- Modify parameter of hearing aids algorithm
- Program external memory to download/update firmware

A more complex OCD with break-point is introduced in a FPGA implementation for firmware development, it has the following functions:

- Memory access
- DSP register access
- Run/Stop/Step
- Instruction break point
- Data break point
- Both OCDs are accessed through JTAG ports.

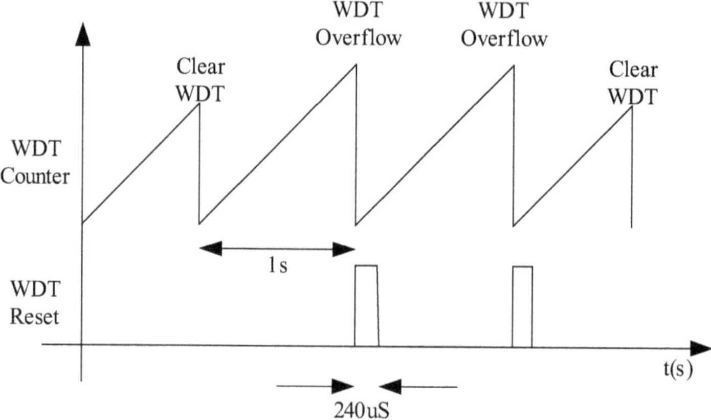

Fig. 17 Watch dog timer timing diagram

Watchdog Timer (WDT)

In hearing aids SoCs, the watchdog timer counts the independent 32 kHz RC clock frequency. When power-on-reset or external reset pin input occurs, the watchdog timer (WDT) is cleared, and then, the WDT starts counting. If the WDT is not cleared in 1 s, a WDT overflow occurs, which generates a reset signal which holds 240 μs. If the timer has not been cleared, 1 s later, a reset signal is generated again. The WDT timing diagram is shown in Fig. 17.

In IDLE mode, the WDT stops working.

2.7 On-Chip Memories

Three SRAM are integrated in hearing aids SoC, one for instruction and two for data. After reset, the boot loader will load the firmware from an off-chip memory and store the firmware into this instruction SRAM.

Two data SRAM allows dual data access at the same time, this feature is very suitable for filter-like algorithm implementation.

Each data SRAM has a direct memory access (DMA) unit, as shown in Fig. 18. DMA0 is used to sample and store data from decimator of ADC to data SRAM periodically. DMA1 is used to load and store data from data SRAM to interpolator of DAC periodically. The DMA dramatically reduce the data access burden of DSP-core.

2.8 Other On-Chip Peripherals

On-chip peripherals include:

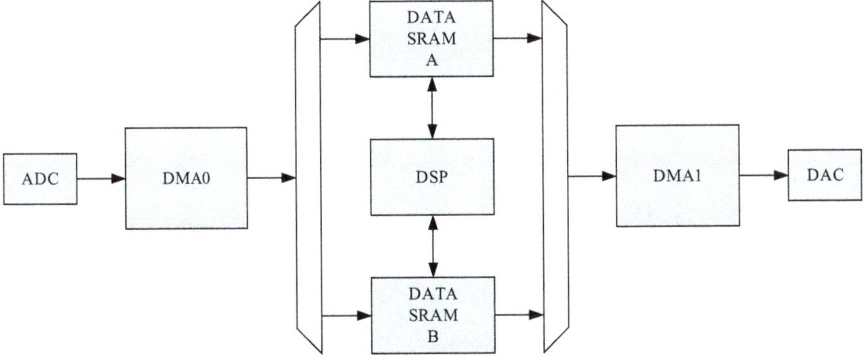

Fig. 18 On-chip data path, with DMA for data access efficiency

- Real Timer Clock (RTC)
- Universal Asynchronous Receiver/Transmitter (UART)
- Timers/Counters
- Liquid Crystal Display (LCD) driver

RTC

The RTC has the following features:

- Real time clock and calendar;
- Calibrating time;
- Pulse output per second, and generating an interrupt.

In IDLE mode, RTC keeps working, and can wake up the system at a configured interval, like 1 day/1 h/1 min/several second. The RC clock is used as a clock source for RTC. The RC clock is not accurate enough for a long term calendar, but as the hearing aids SoC is most likely to work in a battery power application, the RTC can be adjusted when recharging the battery.

Figure 19 show the diagram of RTC. The RTC includes:

- Counter for year/month/day/hour/minute and second;
- A one-second counter to generate a 1-Hz pulse;
- A set of configuration registers.

UART

There are 6 UART serial interfaces on the chip with the following features:

- Fractional baud rate generator to reduce baud rate error;
- Optional 38 kHz carrier wave modulator for remote controller application;

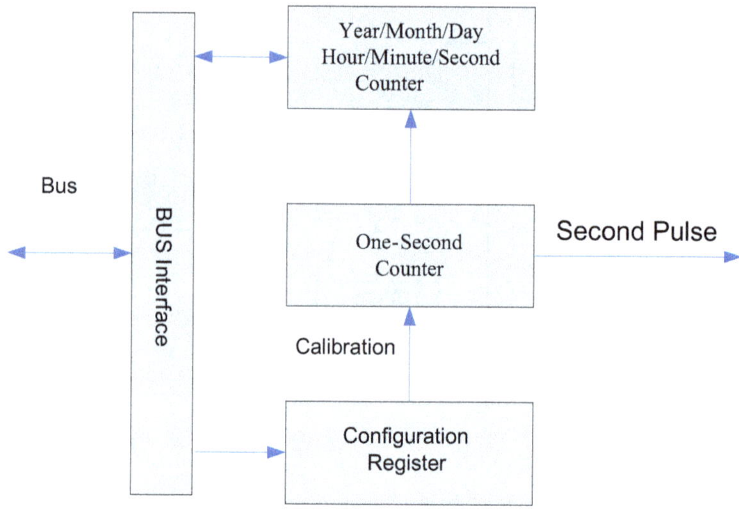

Fig. 19 RTC diagram

- 8/9 bit data;
- Odd/Even/None parity check;
- Multi-drop mode for 485 communication;
- ISO/IEC 7816-3 supportive;
- Asynchronous and full-duplex communication.

Figure 20 shows the diagram of UART module, including:

- Receive shift register with a receive buffer;
- Send shift register with a send buffer;
- A baud-rate generator with fractional feature;
- A set of status and configuration registers;
- A UART FSM unit.

Timer

Two 16-bit timers that have 4 working modes are implemented into hearing aids SoC. Each has 2 compare/capture modules, and 3 configurable output units with 8 output modes for pulse width modulation (PWM) application as shown in the following figure (Fig. 21).

LCD Driver

In some hearing aids equipment, simple LCD panel is used to display status/volume and battery condition. An ultra-low power LCD driver is integrated into the hearing

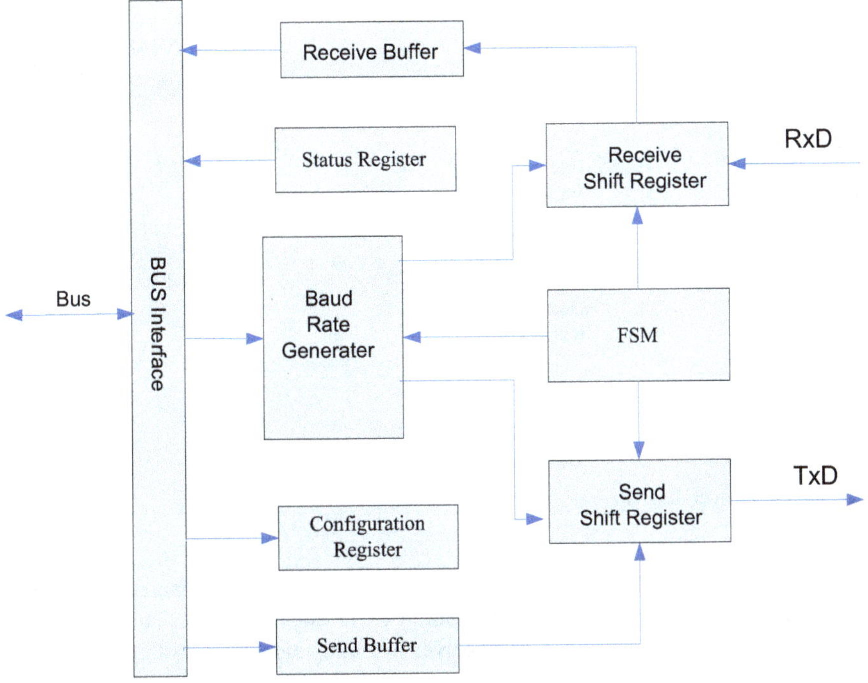

Fig. 20 UART diagram

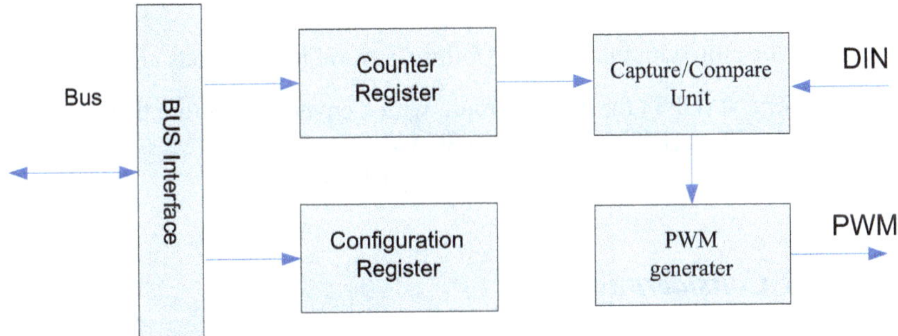

Fig. 21 Timer diagram

aids SoC. It can drive an LCD panel of 4×40 segments. The clock source for the LCD module is RC clock, with a frequency of 32.768 kHz.

Three parameter of LCD module is adjustable for performance/power trade off:

- Though a programmable charge-pump, the driving voltage of the LCD is $3.0\,V \sim 3.3\,V$;
- Scanning frequency is configurable through 64–512 Hz;
- The driving power is adjustable to 4 levels;

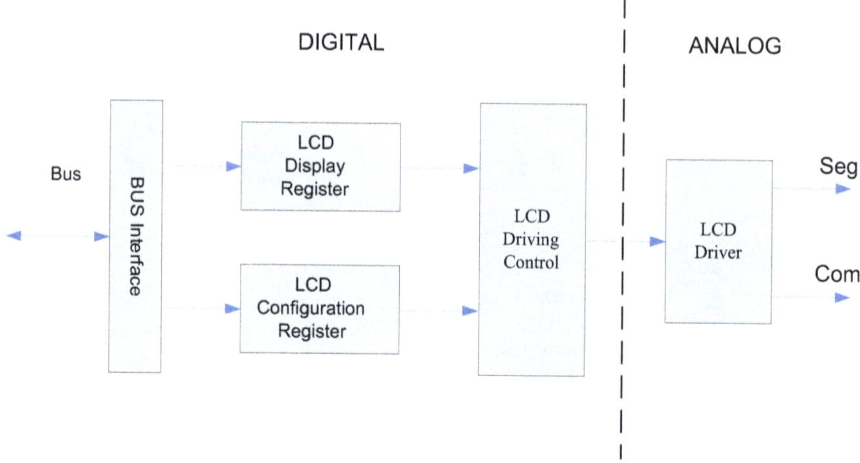

Fig. 22 LCD driver diagram

As shown in Fig. 22, LCD driver consists of digital and analog parts. LCD driving control generate signals based on data in LCD display register and LCD configuration register to control the LCD driver in analog side.

3 Test Consideration

To pick out the chip with manufacture failure, several test methods are introduced:

- Design for test (DFT) for digital circuit, with a coverage of more than 99.9%;
- Build In Self Test (BIST) for on-chip SRAM;
- ADC SNR test in an ATE equipment after package.

3.1 DFT Consideration

Two DFT scan chains are inserted into hearing aids SoC, which need 8 I/Os. In order to save I/O source, most DFT pins share IO with other functions (Table 2).

Interface with Analog/IO and memory are specially designed to fulfill DFT controllable and detectable.

3.2 BIST

BIST is included for all on-chip memory. This feature is enabled at chip probing (CP) test phase for manufacture fault and boot-loader phase for system self-test.

Table 2 Define of DFT IOs

Name	I/O	Description
dft_en	Input	Enable DFT function
dft_clk	Input	Clock of DFT function, active at pos-edge
dft_rstn	Input	DFT reset signal, active low
dft_scan_enable	Input	Scan enable, active high
dft_di[1:0]	Input	DFT scan chain data in
dft_do[1:0]	output	DFT scan chain data out

Fig. 23 BIST test diagram

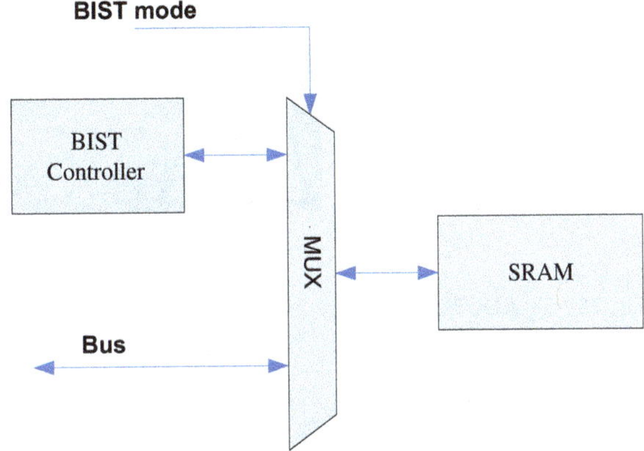

As shown in Fig. 23, in BIST mode, BIST control unit grant the access to on-chip memory to auto-test the memory.

3.3 ADC Test

In CP test phase, bit stream of ADC output and ADC sample clock is multiplexed out to IO pins, an ATE equipment can sample the ADC output and calculate the SNR.

4 Layout, Package, and Results

The hearing aids SoC chip is fabricated in TSMC 0.13 μm mixed-mode technology. The following is a photo of a 100PIN chip (Fig. 24).

The hearing aids SoC has two packages.

- 28 PIN QFN package for hearing device
- 100 PIN LQFP package for ADC test/firmware development/chip verification

For some more complex applications, the hearing aids SoC can also be packaged in 64 pin packages to realize UART communication/RTC feature, etc.

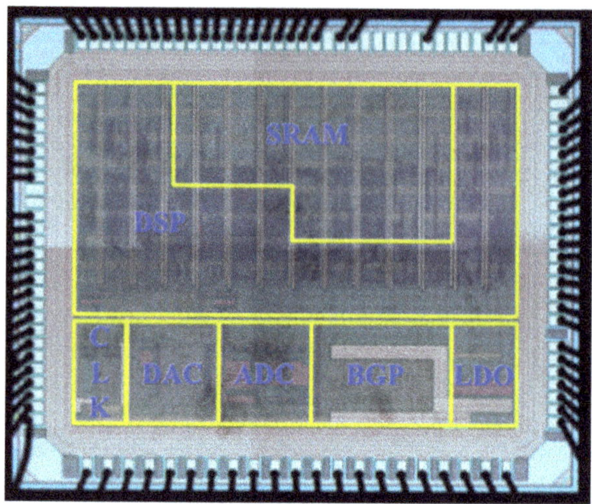

Fig. 24 Die photo of hearing aids SoC

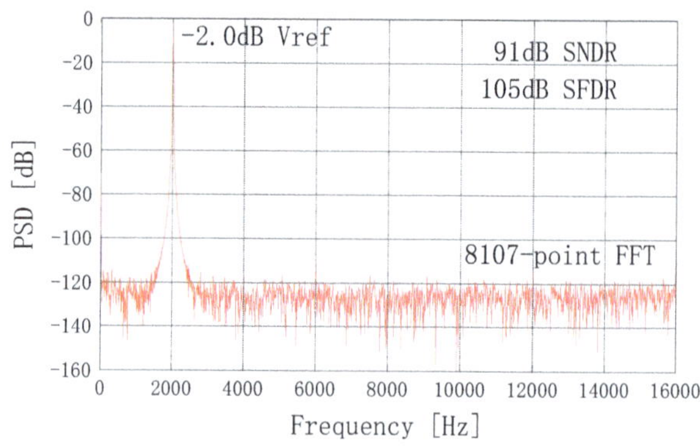

Fig. 25 $\Delta\Sigma$ ADC output spectrum with 1,997 Hz, -1.9 dB Vref sine wave input

The following are some test result of the hearing aids SoC.

Figures 25 and 26 shows the output spectrum of $\Delta\Sigma$ ADC in the SoC.

Figure 27 shows the spectrum of measured 65,536 points Rife-Vincent1-windowed FFTs of the delta-sigma DAC output. The measured DR and peak SNDR in the 20 Hz~10 kHz bandwidth is 90 dB and 85.6 dB respectively.

Table 3 shows some test result of the hearing aids SoC, mostly ADC/DAC performance and power consumption measurement.

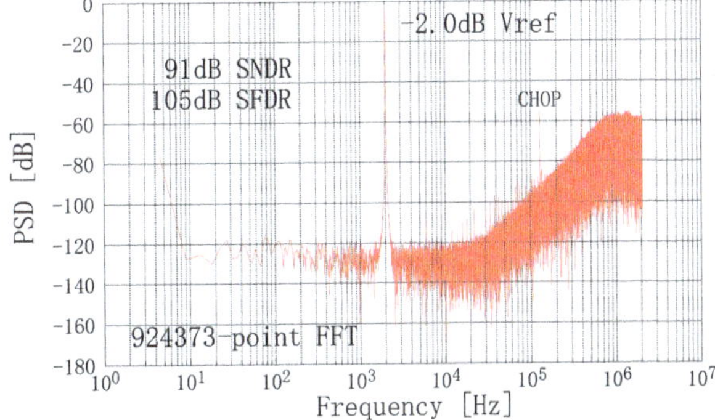

Fig. 26 $\Delta\Sigma$ ADC output spectrum with 1,997 Hz, -1.9 dB Vref sine wave input

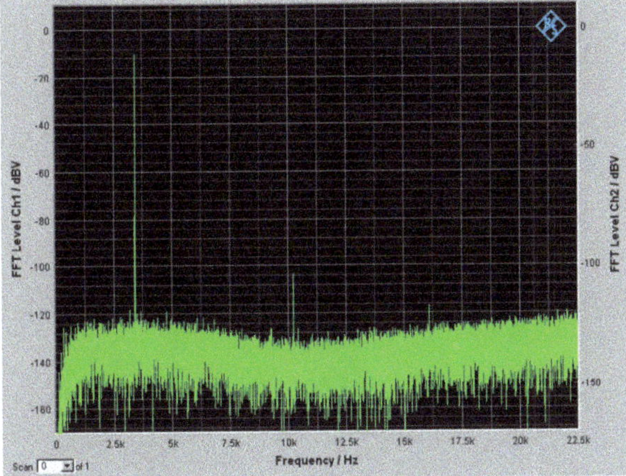

Fig. 27 The DAC ouput with 3.414-kHz -9.6-dBFS sine wave input

Table 3 Test results

	Results
Analog input offset	10 uV
Analog equivalent input noise	20 nV/rtHz
FOM of ADC	<0.4 pJ/Step
Input dynamic range	96 dB (OSR 256)
Output dynamic range	90 dB
PGA range	126 dB
16bit DSP	170 pJ/Instruction
Power consumption	0.7 mW for 8 kHz bandwidth application

References

1. Braida L D, Durlach N I, De Gennaro S V, Peterson P M, Bustamante D K (1982) Review of recent research on multi-band amplitude compression for the hearing impaired. In: Studebaker G A and Bess F H (eds) Monographs in contemporary audiology, The Vanderbilt hearing-aid report, Upper Darby PA, 133–140
2. Byrne D, Dillon H, Ching T, Katsch R, Keidser G (2001) NAL-NL1: Procedure for fitting non-linear hearing aids: characteristics and comparison with other procedure. J Amer Acad Audiol 12:37–51
3. Cudahy E, Levitt H (1994) Digital hearing instruments: A historical perspective. In: Sandlin R E (ed) Understanding digitally programmable hearing aids. Allyn and Bacon, Needham Heights MA, 1–13
4. Dillon H (2001) Hearing aids. Boomerang press, Sydney
5. Fabry D A (1991) Programmable and automatic noise reduction in existing hearing instruments. In: Studebaker G A, Bess F H (eds) The Vanderbilt hearing-aid report II, York Press, Parkton MD, 65–78
6. Holube I, Velde T M (2000) DSP Hearing Instruments. In: Sandlin R E (ed) Textbook of hearing aid amplification. 2nd ed, Singular, San Diego CA, 285–321

SoC for Capsule Endoscope

Hanjun Jiang, Xinkai Chen and Zhihua Wang

1 Introduction to Capsule Endoscope

The wireless capsule endoscope has emerged as a significant supplement to the traditional endoscope for gastrointestinal examination [1] without any anesthesia or insufflations (Fig. 1).

A typical capsule endoscope for the entire digestive tract examination is the Pill-Cam SB capsule manufactured by Given Imaging, which can take 256*256 images at a frame rate of 2 frames per second (fps), and the internal battery can supply the capsule for 8 h [2]. The product by Olympus has the similar performance [3]. The PillCam has an embedded RF transmitter (ZL70081) for one-way data transmission, and the transmitter has a raw data rate of 2.7 Mbps [4]. From the viewpoint of diagnosis, the medical practitioners would like to have higher image resolution and higher frame rate. The key challenges of capsule endoscope IC design lies on lowering the power consumption, maintaining high data rate, and integrating the components into the silicon.

With the limited communication bandwidth, a higher frame rate can still be achieved by using the technique of lossless/near-lossless image compression [5, 6]. When adopting the image compression technique, the communication bit-error-rate (BER) has to be lowered to around 10^{-10} to prevent error propagation during image data decompression. However, this level of BER is hard to reach by using one way data transmitting with affordable channel coding. Instead, a bi-directional communication which supports automatic repeat request (ARQ) is required to maintain the BER adequately low. On the other hand, a bi-directional communication is also required to configure/adjust the working status of the capsule.

H. Jiang (✉) · Z. Wang
Institute of Microelectronics, Tsinghua University,
Beijing, China
e-mail: jianghanjun@tsinghua.edu.cn

X. Chen
Ecore Technologies Ltd., Beijing, China

N. N. Tan et al. (eds.), *Ultra-Low Power Integrated Circuit Design,*
Analog Circuits and Signal Processing 85, DOI 10.1007/978-1-4419-9973-3_10,
© Springer Science+Business Media New York 2014

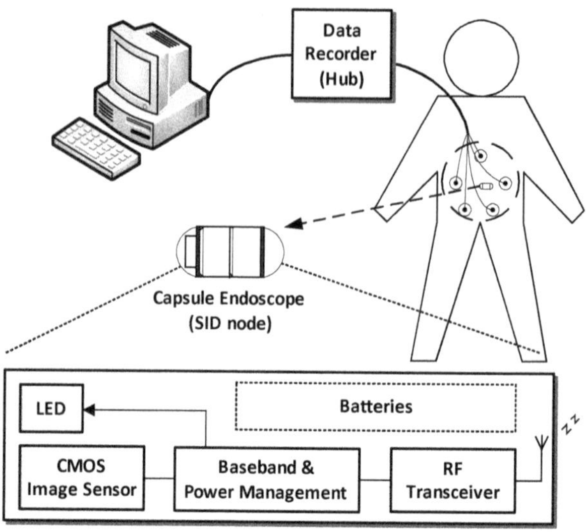

Fig. 1 Wireless capsule endoscope system

In this section, an application specific SoC with bidirectional communication for wireless capsule endoscope is discussed. The SoC has an integrated ultra-low-power transceiver featured asymmetrical data rate (3 Mbps transmitting and 64 kbps receiving) as shown in Chap. 6. A dedicated ARQ scheme is designed to maintain an adequately low BER for data transmission. A near-lossless image compressor is integrated to improve the system efficiency. A capsule endoscope built using this SoC can transmit 480 × 480 pixel images at a frame rate of up to 3 fps. The bi-directional link also renders the capability of re-configuring the capsule working inside the human body.

2 System and SoC Architecture

A capsule endoscope can be built with only two silicon chips, one is the designed SoC, and the other is an image sensor. As shown in Fig. 2, the SoC is composed of four key function blocks: RF transceiver, microcontroller, image processor, and the power management unit.

The detailed functional block diagram of this SoC is shown in Fig. 3. The required external components to form a capsule endoscope are also shown Fig. 3. If we don't count the flash LEDs and the antenna, the SoC only need 7 external components (6 capacitors for voltage regulation and decoupling, and one 24 MHz crystal for frequency reference).

The architecture of the RF transceiver is depicted at the right-up corner of Fig. 3, which has be explained in Chap. 6. The microcontroller (MCU), the media access controller (MAC), the image processor (JPEG-LS Encoder and FIFO), and other digital functions such as the image sensor controller and test-mode controller are designed and synthesized as one circuit block, and will be discussed in Sect. IV.

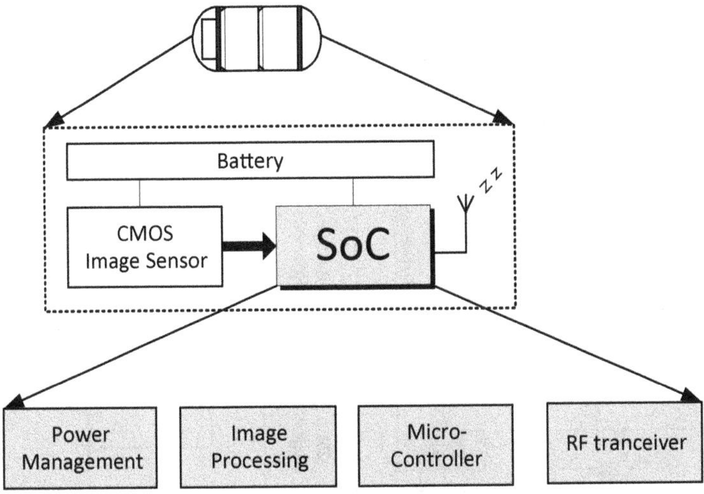

Fig. 2 System architecture of the capsule endoscope

The power management unit is composed of three programmable low-dropout (LDO) linear regulators supplying other function blocks, and one programmable boost charge pump [5] for LED driving. Other necessary blocks include a band-gap circuit for voltage and current references, a crystal oscillator for precise frequency reference, a 2 µA RC relaxation oscillator for status control, a power-on reset (POR) block, and a remotely controlled power switch based on wireless energy recovery [5].

3   Digital Core

The whole SoC has an ultra-low-power digital core, which takes care of system control, image data processing and the communication protocol. The power management unit and the RF transceiver are controlled by the digital core as well. The digital core architecture is based on the Wishbone bus. Three master modules (including the MCU, I²C controller for test and the watchdog timer) and six slave modules (the MAC controller, image processor, and etc.) are connected to the bus as shown in Fig. 4.

The MCU is the most significant part of the digital core. It is actually a dedicated low-power control unit with very limited instructions. The high-level MAC protocol is implemented in the program of the MCU, while the real-time image compression and the low-level MAC protocol are accelerated by the specific hardware. This architecture provides adequate function flexibility with high energy efficiency. The techniques of clock-gating and adjustable digital supply voltage are applied to reduce energy dissipation further.

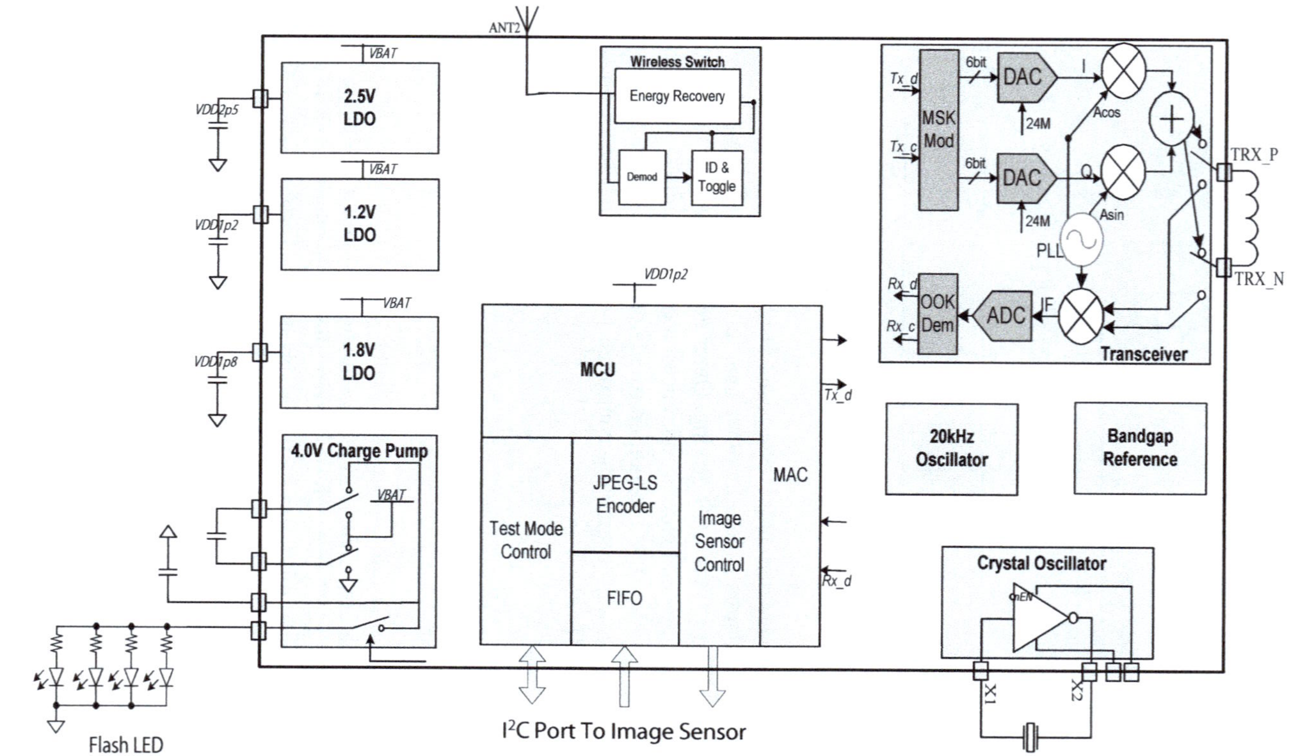

Fig. 3 Block diagram of the SoC (inside the thick line) and the external components

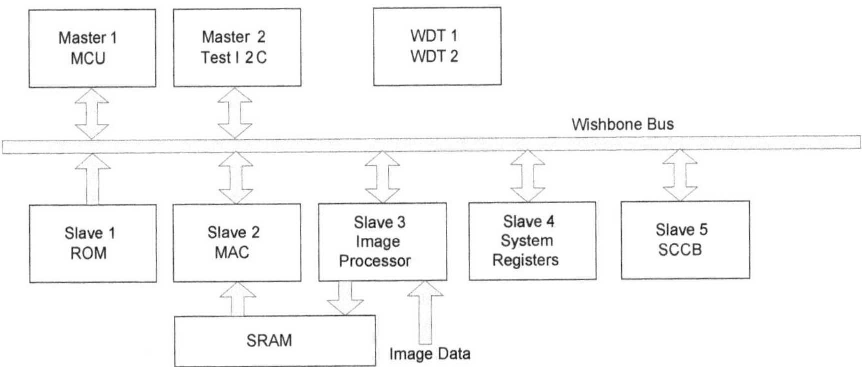

Fig. 4 Architecture of the digital core

The image processor is a dedicated block which can provide near-lossless image data compression which helps to reduce the data amount to transmit, and consequently to improve the system power efficiency. In this design, the data compression is optional, and can be bypassed when needed.

Considering the channel noise, the WABN hub in the application system with a receiving sensitivity of -90 dBm can achieve a raw receiving BER of around 10^{-3}, which is definitely not acceptable for image data transmission. Channel coding is required for the system. In the transmission path, the transmitted data is protected by the cyclic redundancy check (CRC) code and the Reed-Solomon (RS) code, and whitened before transmission. The receiving path has 4B5B decoding, de-whitening, RS decoding, and CRC decoding, correspondingly.

The RS code is chosen to improve the BER from 10^{-3} to 10^{-7}. A BER of 10^{-7} is acceptable when transmitting un-compressed images with 1/4 Mega pixels (480×480). When transmitting compressed images for a higher frame rate, ARQ is utilized to improve the BER further based on the bidirectional communication. Note that the receiver in this SoC consumes higher power than the transmitter, the ARQ scheme is designed such that the receiver is powered on only for the minimal time. For this consideration, data re-transmission in used only on a frame base, and the capsule SoC will accept re-transmission request from the external data logger only at the end of transmitting each image frame. The implemented ARQ can support a capsule endoscope with 3 fps frame rate even for 512*512 images (Fig. 5).

Ultra-low-power digital circuit design techniques have been adopted to reduce the power consumption of the SoC digital circuit including the image compressor.

3.1 Clock Gating

Clock gating is extensively used as an intrinsic way to implement logic functions and to save power. The flip-flop toggling statistics is gathered at the RTL level of the design and provide early feedback on the effectiveness of clock gating in each func-

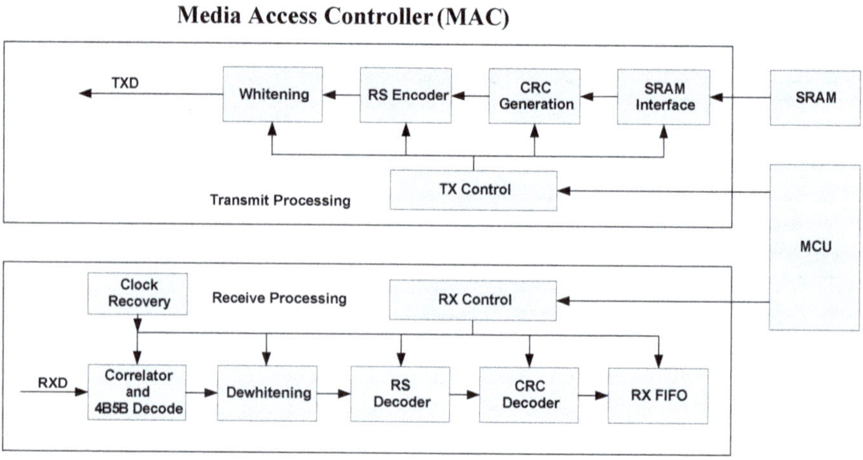

Fig. 5 Media access controller

tion units. Power consumption is re-estimated with actual parasitic extraction using commercial tools at physical implementation level. The simulation results have shown the good correlation of power estimation between RTL and transistor levels.

3.2 *Voltage Scaling*

Voltage Scaling is a very efficient mean to reduce the dynamic power since the power dissipation is proportional to VDD2. In this SoC power management block, an integrated LDO is applied to provide the supply voltage for the digital core. The LDO output voltage can be scaled by tuning the LDO feedback network, which is done by the MCU. In the full work mode, the supply voltage is scaled to the predefined value in different phases. The test result shows that the supply voltage can be safely reduced to 1.1 V at image compression phase when 24 MHz clock frequency is applied. To reduce the static power, the supply voltage can be lowered even further and the oscillator can be shut down when the system is disabled.

4 Image Compression

A low complexity, high performance lossless/near-lossless image compressor is implemented in the SoC. The flow chart of the entire image compression algorithm is shown in Fig. 6. It is mainly composed of an image filter followed by a standard JPEG-LS encoder.

The image compression module is dedicatedly designed for a CMOS image sensor with Bayer color filter array (CFA). For image data in the CFA format, there is only

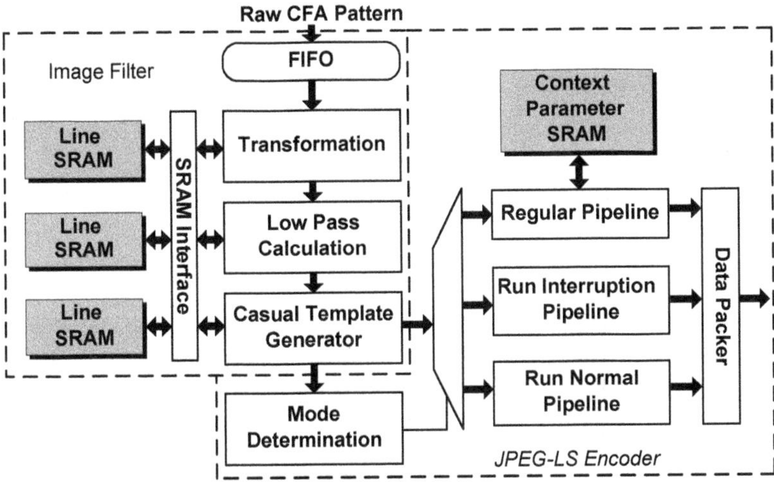

Fig. 6 Block diagram of image compression module

one color component in each pixel and the other two color components for the given pixel need to be interpolated using neighboring pixel information to generate a full color image. In most applications, the full color image is compressed before storage or transmission which results in three times data amount and processing time compared to the CFA image. This data redundancy can be removed by moving the compression stage before color interpolation. By compressing the CFA pattern before the interpolation, more pertinent information is retained, allowing a higher compression rate while maintaining the image quality [7]. Furthermore, this scheme could greatly save the hardware cost such as the on-chip memory storage and the communication bandwidth for wireless data transmission. For these reasons, the Bayer CFA pattern is directly compressed and transmitted in the proposed image compression module. The VLSI architecture of the image compression module is shown in Fig. 6. It is composed of an image filter and a JPEG-LS encoder which will be described in detail next.

4.1 Hardware-Oriented Image Filter

A characteristic associated with the Bayer CFA pattern is that for a given pixel the neighboring pixels always belong to different color components, as shown in Fig. 7. The correlation between two neighboring pixels is very low and the influence of high spatial frequencies is extremely notable. Therefore, algorithms based on prediction such as DPCM, or based on transformation such as DWT, suffer from a severe degradation on the compression performance. In this SoC, an image filter algorithm is introduced to alleviate this problem.

For the purpose of real time data processing and low memory requirement, this SoC utilizes a real-time image filter algorithm, in which the data are firstly

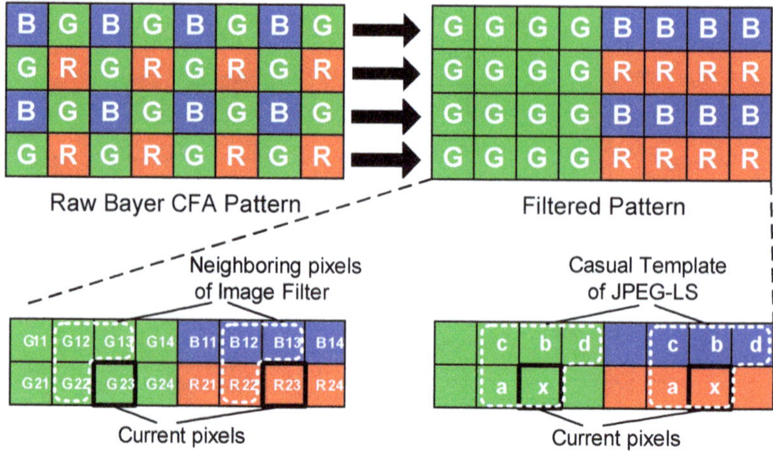

Fig. 7 Proposed image filtering process and casual template generation of JPEG-LS algorithm

transformed and then low-pass filtered directly in RGB space to depress the high spatial frequencies. Afterwards, the casual template is generated and sent to the JPEG-LS encoder for compression, as shown in Fig. 6. As illustrated in Fig. 7, the G component and the B&R components are separated into two rectangular arrays by transformation operation. The high spatial frequencies between neighboring pixels due to different color component belonging are greatly decreased, which helps the following JPEG-LS encoder helpful to obtain a relative high compression rate for. In this method, B and R components are processed together as a singlecolor component in the filtering procedure. To make a good trade-off between the cost and the performance, two-color-components filter method is used in this design as described above. The filtering algorithm can be implemented in two steps.

- [Step 1] Transformation: G components and B&R components are separated into two rectangular arrays.

$$\begin{cases} X = x \\ Y = y + \left((-1)^{x+y} + 1 \right) \cdot N / 4 \end{cases} \tag{1}$$

where (x, y) is the coordination of the current pixel, while (X, Y) indicates the coordination after transformation. N indicates the width of the image.

- [Step 2] Low pass filtering: the current pixel is averaged with the neighboring pixels within both horizontal and vertical direction.

$$P'_{(X,Y)} = \frac{1}{4\left(\left(P_{(X,Y)} + P'_{(X-1,Y-1)} + 1 \right) + \left(P'_{(X-1,Y)} + P'_{(X,Y-1)} + 1 \right) \right)} \tag{2}$$

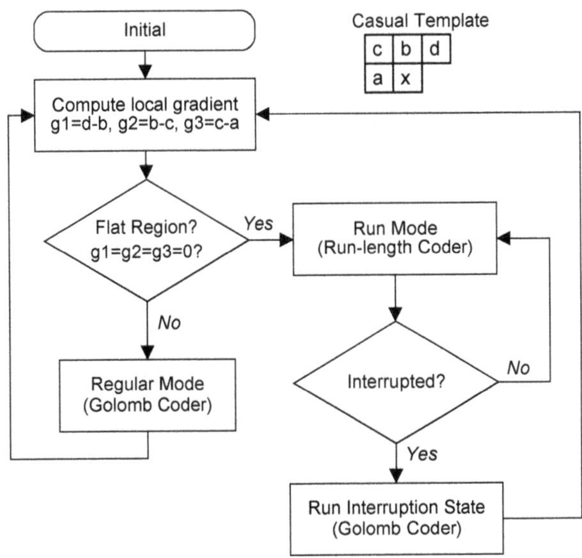

Fig. 8 Simplified flow diagram of JPEG-LS algorithm

where P indicates the original value of the current pixel, P' indicates the filtered value.

For the current pixel, the filtering algorithm only utilizes the available past neighboring pixels within casual template of the JPEG-LS algorithm. By doing so, it helps to achieve real-time operation with small memory storage, and also facilitates the causal template generation of JPEG-LS. Due to the limited bit width, the division operation in (1) will introduce error that no more than 2 after reconstruction.

4.2 JPEG-LS Encoder

JPEG-LS is an established standard for lossless and near-lossless compression of still images [8]. It provides both the highest lossless compression ratio and the fastest compression speed for medical images. The main compression techniques of JPEG-LS [9] can be summarized as follows:

- Context Modeling: It performs local gradient computation, gradient quantization and quantized gradient merging.
- Context-based statistic: Context parameters $A[Q]$, $B[Q]$, $C[Q]$ and $N[Q]$ are iterated for each context model.
- Predictor: It is composed of fixed predictor and adaptive correction. The prediction residual is calculated for coding.
- Run-length coding: it is used for the smooth local region.
- Limited-length Golomb coding: it is used for prediction residual coding.

Figure 8 shows the simplified flow chart, in which a simple mode selection strategy is adopted. Based on the local gradient ($g1$, $g2$, $g3$) computation, each pixel

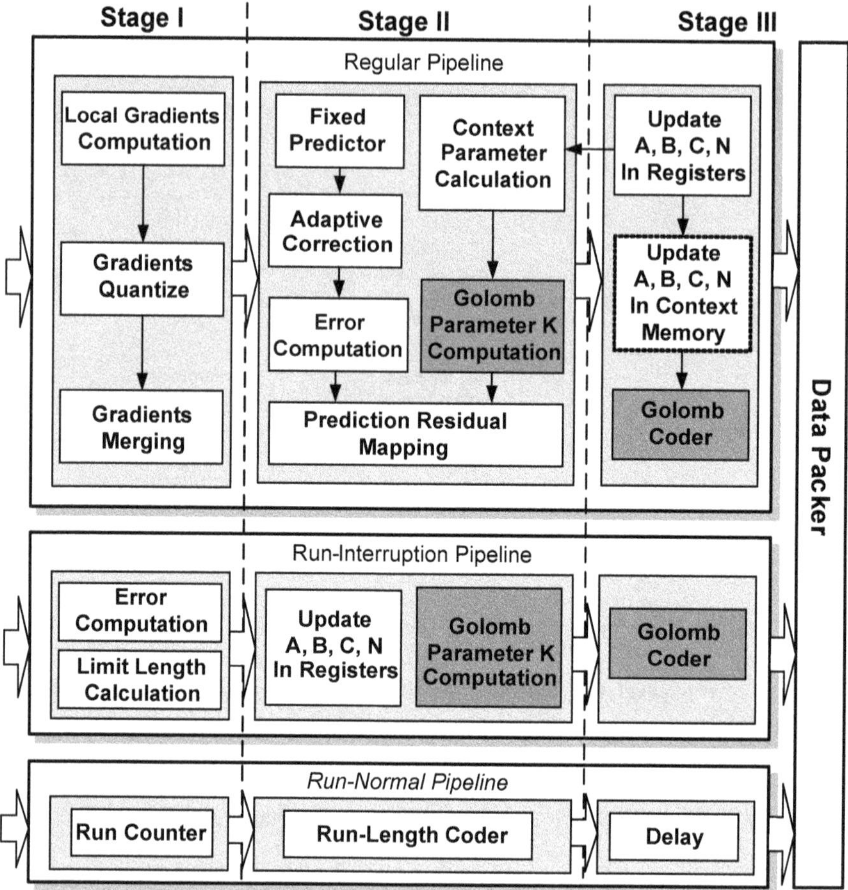

Fig. 9 Data path of JPEG-LS with three pipelines

would be encoded either in the regular mode or in the run mode by Golomb coder or run-length coder. The computation in the regular mode and the run interruption state contributes the most to the computation complexity of the JPEG-LS algorithm. The mapped prediction residual computation in the regular mode is always the most time-consuming part. Moreover, the update of context parameters $A[Q]$, $B[Q]$, $C[Q]$, $N[Q]$ requires extra accessing time of memory. Thus, appropriate hardware architecture has to be considered to meet the real-time processing requirement.

To achieve an area and power efficient VLSI implementation of the JPEG-LS encoder, the designed data path is optimized targeting the minimum resource utilization and power consumption. In our design, three parallel three-stage pipelined data paths are implemented. The three pipelines take care of the regular mode, the run mode and the run interruption state, respectively, as shown in Fig. 9. Only one of the three pipelines is activated according to the mode determination, which avoids the unnecessary computation of the other two pipelines for each pixel. The simulation

Fig. 10 Die photo of the implemented SoC

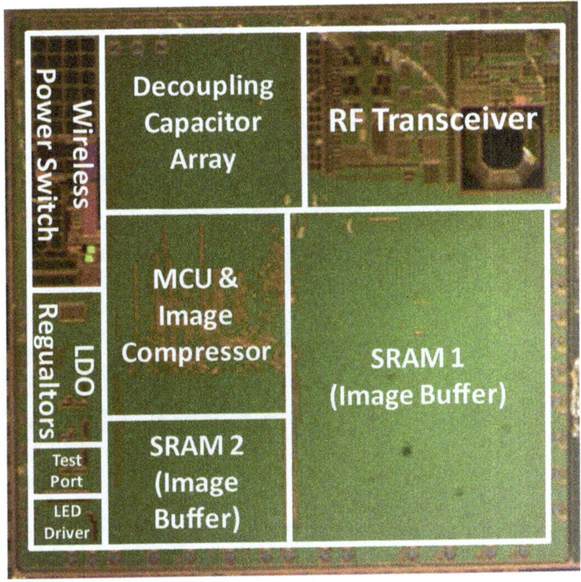

result of power consumption of data path shows that separated pipelines for different modes would result in an average of 90 % power reduction in the run mode. This is expected when considering the low computation complexity of the run pipeline. With this method, unnecessary switch activities are effectively reduced, especially when images are smooth. In addition, resource sharing is used among pipelines for common computation such as Golomb parameter k computation in the second stage and Golomb coder in the third stage, as shown in the dark gray color in Fig. 9.

A two-level hierarchy memory access method is proposed to eliminate unnecessary memory accesses in the process of context parameters updating. The experiments have shown that the proposed method results in an average of 52 % context memory accesses reduction which means considerable power reduction.

5 Implementation Results

The SoC has been implemented in 0.18 μm CMOS technology.

The digital core has 30 k equivalent gates, and 94 kBytes SRAM for image buffering. The die photo of the SoC is shown in Fig. 10. It occupies a die area of 13.3 mm² (3.7 × 3.6 mm). A capsule endoscope with a diameter of 11 mm and a length of 26.5 mm has been built with the presented SoC as shown in Fig. 11.

The SoC can work at a voltage down to 2.5 V. The power consumption of the SoC has been checked. The MCU consumes about 200 μA current from a 1.2 V power supply (from the on-chip regulator) when clocked at 24 MHz. The image compressor consumes about 900 μA current from the 1.2 V supply. The MSK transmitter

Fig. 11 Capsule endoscope prototype

Table 1 Capsule endoscope SoC performance summary

Parameters		This work
Supply voltage		2.5 V ~ 3.3 V
External components #		7
TX	Type of RF link	Bi-directional
	Bit rate	3 Mbps
	Power consumption	3.9 mW
RX	Bit rate	64 kbps
	Power consumption	12 mW
MCU power consumption		240 µW
Image compressor power		1.1 mW
Image resolution		$480 \times 480 / 240 \times 240$
Frame rate		Up to 3 fps @ 480×480 w/ compression
Technology		0.18 um CMOS
Die area		13.3 mm²

(including all the functional blocks for TX) consumes a total power for 3.9 mW from the 2.5, 1.8 and 1.2 V supplies, while the OOK receiver consumes 12 mW.

Figure 12 shows some typical gastrointestinal images taken by the prototype capsule endoscope.

The performance of this SoC is summarized in Table 1.

In summary, an ultra-low-power SoC has designed for the application of wireless capsule endoscope. The SoC has a RF transceiver working at the 400 MHz UHF band, with a 3 Mbps MSK transmitter and a 64 kbps OOK receiver, enabling the feature of bidirectional communication between the capsule and the external data logger. The MSK transmitter consumes only 3.9 mW power in total. A 1.2 V MCU with a dedicated image compressor and a media access controller is implemented to improve the system efficiency. With the SoC, a capsule endoscope can transmit 480×480 images at a frame rate of 3 fps. A prototype system has been developed to validate the SoC performance.

Fig. 12 Typical gastrointestinal images taken by the prototype capsule endoscope

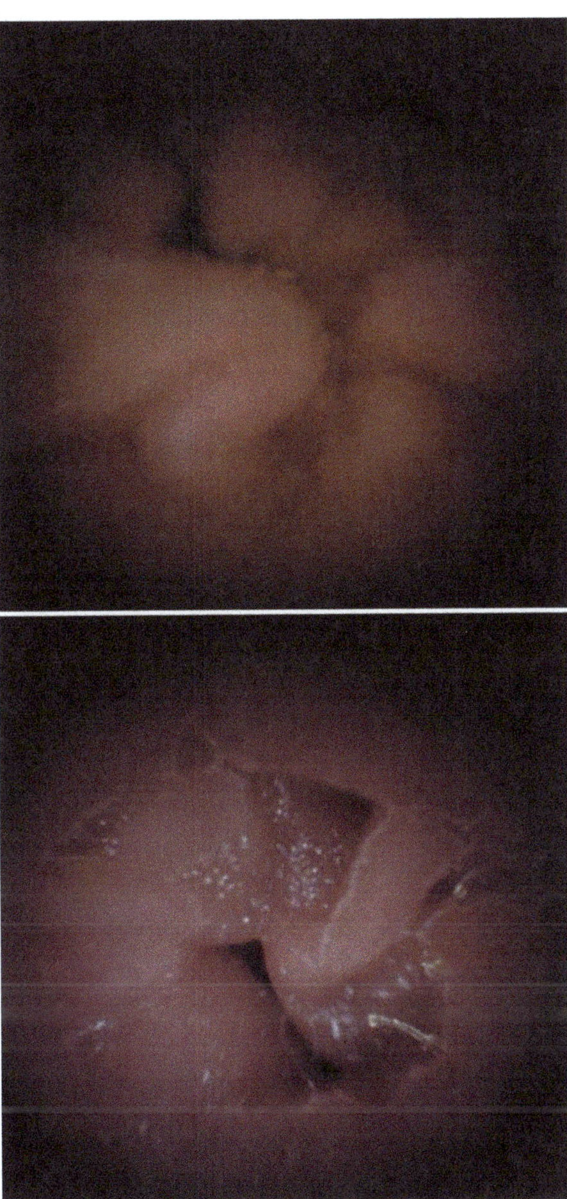

References

1. G. Iddan, G. Meron, A. Glukhovsky, et al. "Wireless capsule endoscopy," Nature, vol 405, pp. 417–418, May 25, 2000.
2. Given Imaging, Overview of Products, available online: http://www.givenimaging.com/en-us/Patients/Pages/pageSmallBowel.aspx

3. Olympus, "Olympus Launches High-resolution Capsule Endoscope in Europe," available online: http://www.olympus-global.com/en/news/2005b/nr051013capsle.html
4. P. Bradley, "RF Integrated Circuits for Medical Implants: Meeting the Challenge of Ultra Low-Power," available online: http://www.cmoset.com/uploads/Peter_Bradley.pdf
5. X. Chen, X. Zhang, L. Zhang, X. Li, N. Qi, H. Jiang, Z. Wang, "A wireless capsule endoscope system with low-power controlling and processing asic," IEEE Trans. on BioCAS, vol. 3, no. 1, pp. 11–22, Feb. 2009
6. X. Xie, G. Li, X. Chen, X. Li, and Z. Wang, "A low-power digital ic design inside the wireless endoscopic capsule," IEEE Journal of Solid-State Circuits, vol. 41, no. 11, pp. 2390–2400, Nov. 2006
7. C. C. Koh, J. Mukherjee, and S. K. Mitra, "New Efficient Methods of Image Compression in Digital Cameras with Color Filter Array," IEEE Trans. Consumer Electronics, vol. 49, No. 4, pp. 1448–1456, Nov. 2003
8. Information Technology-Lossless and near-lossless compression of continuous-tone images-Baseline. International Telecommunication Union (ITU-T Recommendation T.87). ISO/IEC 14495-1, 1998
9. M. Weinberger, G. Seroussi, G. Sapiro, "The LOCO-I lossless image compression algorithm: principles and standardization into JPEG-LS," IEEE Trans. Image Processing, vol. 9, no. 8, pp. 1309–1324, Aug. 2000.

GPSR Compliance
The European Union's (EU) General Product Safety Regulation (GPSR) is a set
of rules that requires consumer products to be safe and our obligations to
ensure this.

If you have any concerns about our products, you can contact us on

ProductSafety@springernature.com

In case Publisher is established outside the EU, the EU authorized
representative is:

Springer Nature Customer Service Center GmbH
Europaplatz 3
69115 Heidelberg, Germany